Zoology: Study of The Animal Kingdom

Zoology: Study of The Animal Kingdom

Kate Porter

Larsen & Keller
www.larsen-keller.com

Zoology: Study of The Animal Kingdom
Kate Porter
ISBN: 978-1-64172-510-1 (Hardback)

▤ Larsen & Keller

Published by Larsen and Keller Education,
5 Penn Plaza,
19th Floor,
New York, NY 10001, USA

Cataloging-in-Publication Data

Zoology : study of the animal kingdom / Kate Porter.
    p. cm.
Includes bibliographical references and index.
ISBN 978-1-64172-510-1
1. Zoology. 2. Animals. 3. Natural history. I. Porter, Kate.
QL45.2 .Z66 2020
590--dc23

For more information regarding Larsen and Keller Education and its products, please visit the publisher's website www.larsen-keller.com

Table of Contents

Preface

This book aims to help a broader range of students by exploring a wide variety of significant topics related to this discipline. It will help students in achieving a higher level of understanding of the subject and excel in their respective fields. This book would not have been possible without the unwavered support of my senior professors who took out the time to provide me feedback and help me with the process. I would also like to thank my family for their patience and support.

The biological branch which deals with the study of evolutionary, structural and behavioral characteristics of animals is referred to as zoology. It also focuses on the classification, habits and distribution of living and extinct animals as well as their interaction with their ecosystems. There are various sub-disciplines of zoology such as zoography, comparative anatomy, behavioral ecology, ethology, vertebrate zoology, soil zoology, invertebrate zoology and animal physiology. Zoography is an applied science which deals with animals and their habitats. Comparative anatomy is involved in the study of the structure of animals. Behavioral ecology is concerned with the evolutionary basis for animal behavior due to ecological pressures. Zoology also branches out into various taxonomically oriented disciplines such as mammalogy, herpetology, ornithology, entomology and biological anthropology. This textbook attempts to understand the multiple branches that fall under the discipline of zoology and how such concepts have practical applications. It provides comprehensive insights into this field. This book is a complete source of knowledge on the present status of this important field.

A brief overview of the book contents is provided below:

Chapter – Introduction

The branch of biology that is concerned with the study of the animal kingdom including the evolution, embryology, habits, structure and classification of animals is referred to as zoology. This is an introductory chapter which will introduce briefly all the significant aspects of zoology.

Chapter – Branches of Zoology

Herpetology, mammalogy, ornithology, entomology, ichthyology, primatology and anthrozoology are some of the branches of zoology. All these branches of zoology have been thoroughly discussed in this chapter to give a better understanding of the subject.

Chapter – Vertebrate Zoology

Vertebrate zoology is a sub-discipline of biology which studies animals that have vertebrae such as amphibians, fish, mammals, birds and reptiles. The chapter closely examines the key concepts of vertebrate zoology as well as these vertebrates to provide an extensive understanding of the subject.

Chapter – Invertebrate Zoology

The sub-discipline of zoology which studies the animals without a backbone is known as invertebrate zoology. Some of the phyla that are studied within invertebrate zoology are arthropoda, mollusca, annelida and nematoda. All the diverse aspects related to these phyla within invertebrate zoology have been carefully analyzed in this chapter.

Chapter – Animal Physiology and Anatomy

The study of the life-supporting properties, processes and functions of animals is known as animal physiology. The branch of biology which studies the structure of animals as well as their parts is known as animal anatomy. The topics elaborated in this chapter will help in gaining a better perspective about the major areas of study within animal physiology and anatomy.

Chapter – Animal Behavior

The scientific study of animal behavior takes place under the domain of ethology. It views their behavior as an evolutionary adaptive trait. This chapter has been carefully written to provide an easy understanding of the varied facets of animal behavior such as their learned behavior, adjunctive behavior, social behavior, territorial behavior and sexual behavior.

Kate Porter

1

Introduction

The branch of biology that is concerned with the study of the animal kingdom including the evolution, embryology, habits, structure and classification of animals is referred to as zoology. This is an introductory chapter which will introduce briefly all the significant aspects of zoology.

ZOOLOGY

Zoology is the branch of biology that studies the members of the animal kingdom and animal life in general. It includes both the inquiry into individual animals and their constituent parts, even to the molecular level, and the inquiry into animal populations, entire faunas, and the relationships of animals to each other, to plants, and to the nonliving environment. Though this wide range of studies results in some isolation of specialties within zoology, the conceptual integration in the contemporary study of living things that has occurred in recent years emphasizes the structural and functional unity of life rather than its diversity.

Prehistoric man's survival as a hunter defined his relation to other animals, which were a source of food and danger. As man's cultural heritage developed, animals were variously incorporated into man's folklore and philosophical awareness as fellow living creatures. Domestication of animals forced man to take a systematic and measured view of animal life, especially after urbanization necessitated a constant and large supply of animal products.

Study of animal life by the ancient Greeks became more rational, if not yet scientific, in the modern sense, after the cause of disease—until then thought to be demons—was postulated by Hippocrates to result from a lack of harmonious functioning of body parts. The systematic study of animals was encouraged by Aristotle's extensive descriptions of living things, his work reflecting the Greek concept of order in nature and attributing to nature an idealized rigidity.

In Roman times Pliny brought together in 37 volumes a treatise, Historia naturalis, that was an encyclopaedic compilation of both myth and fact regarding celestial bodies, geography, animals and plants, metals, and stone. Volumes VII to XI concern zoology; volume VIII, which deals with the land animals, begins with the largest one, the elephant. Although Pliny's approach was naïve, his scholarly effort had a profound and lasting influence as an authoritative work.

Zoology continued in the Aristotelian tradition for many centuries in the Mediterranean region and by the Middle Ages, in Europe, it had accumulated considerable folklore, superstition, and moral symbolisms, which were added to otherwise objective information about animals. Gradually, much of this misinformation was sifted out: naturalists became more critical as they compared directly observed animal life in Europe with that described in ancient texts. The use of the printing press in the 15th century made possible an accurate transmission of information. Moreover, mechanistic views of life processes (i.e., that physical processes depending on cause and effect can apply to animate forms) provided a hopeful method for analyzing animal functions; for example, the mechanics of hydraulic systems were part of William Harvey's argument for the circulation of the blood—although Harvey remained thoroughly Aristotelian in outlook. In the 18th century, zoology passed through reforms provided by both the system of nomenclature of Carolus Linnaeus and the comprehensive works on natural history by Georges-Louis Leclerc de Buffon; to these were added the contributions to comparative anatomy by Georges Cuvier in the early 19th century.

Physiological functions, such as digestion, excretion, and respiration, were easily observed in many animals, though they were not as critically analyzed as was blood circulation.

Following the introduction of the word cell in the 17th century and microscopic observation of these structures throughout the 18th century, the cell was incisively defined as the common structural unit of living things in 1839 by two Germans: Matthias Schleiden and Theodor Schwann. In the meanwhile, as the science of chemistry developed, it was inevitably extended to an analysis of animate systems. In the middle of the 18th century the French physicist René Antoine Ferchault de Réaumer demonstrated that the fermenting action of stomach juices is a chemical process. And in the mid-19th century the French physician and physiologist Claude Bernard drew upon both the cell theory and knowledge of chemistry to develop the concept of the stability of the internal bodily environment, now called homeostasis.

The cell concept influenced many biological disciplines, including that of embryology, in which cells are important in determining the way in which a fertilized egg develops into a new organism. The unfolding of these events—called epigenesis by Harvey—was described by various workers, notably the German-trained comparative embryologist Karl von Baer, who was the first to observe a mammalian egg within an ovary. Another German-trained embryologist, Christian Heinrich Pander, introduced in 1817 the concept of germ, or primordial, tissue layers into embryology.

In the latter part of the 19th century, improved microscopy and better staining techniques using aniline dyes, such as hematoxylin, provided further impetus to the study of internal cellular structure.

By this time Darwin had made necessary a complete revision of man's view of nature with his theory that biological changes in species occur through the process of natural selection. The theory of evolution—that organisms are continuously evolving into highly adapted forms—required the rejection of the static view that all species are especially created and upset the Linnaean concept of species types. Darwin recognized that the principles of heredity must be known to understand how evolution works; but, even though the concept of hereditary factors had by then been formulated by Mendel, Darwin never heard of his work, which was essentially lost until its rediscovery in 1900.

Genetics has developed in the 20th century and now is essential to many diverse biological disciplines. The discovery of the gene as a controlling hereditary factor for all forms of life has been a major accomplishment of modern biology. There has also emerged clearer understanding of the interaction of organisms with their environment. Such ecological studies help not only to show the interdependence of the three great groups of organisms—plants, as producers; animals, as consumers; and fungi and many bacteria, as decomposers—but they also provide information essential to man's control of the environment and, ultimately, to his survival on Earth. Closely related to this study of ecology are inquiries into animal Behavior, or ethology. Such studies are often cross disciplinary in that ecology, physiology, genetics, development, and evolution are combined as man attempts to understand why an organism behaves as it does. This approach now receives substantial attention because it seems to provide useful insight into man's biological heritage—that is, the historical origin of man from nonhuman forms.

The emergence of animal biology has had two particular effects on classical zoology. First, and somewhat paradoxically, there has been a reduced emphasis on zoology as a distinct subject of scientific study; for example, workers think of themselves as geneticists, ecologists, or physiologists who study animal rather than plant material. They often choose a problem congenial to their intellectual tastes, regarding the organism used as important only to the extent that it provides favourable experimental material. Current emphasis is, therefore, slanted toward the solution of general biological problems; contemporary zoology thus is to a great extent the sum total of that work done by biologists pursuing research on animal material.

Second, there is an increasing emphasis on a conceptual approach to the life sciences. This has resulted from the concepts that emerged in the late 19th and early 20th centuries: the cell theory; natural selection and evolution; the constancy of the internal environment; the basic similarity of genetic material in all living organisms; and the flow of matter and energy through ecosystems. The lives of microbes, plants, and animals now are approached using theoretical models as guides rather than by following the often restricted empiricism of earlier times. This is particularly true in molecular studies, in which the integration of biology with chemistry allows the techniques and quantitative emphases of the physical sciences to be used effectively to analyze living systems.

Areas of Study

Although it is still useful to recognize many disciplines in animal biology—e.g., anatomy or morphology; biochemistry and molecular biology; cell biology; developmental studies (embryology); ecology; ethology; evolution; genetics; physiology; and systematics—the research frontiers occur as often at the interfaces of two or more of these areas as within any given one.

Anatomy or Morphology

Descriptions of external form and internal organization are among the earliest records available regarding the systematic study of animals. Aristotle was an indefatigable collector and dissector of animals. He found differing degrees of structural complexity, which he described with regard to ways of living, habits, and body parts. Although Aristotle had no formal system of classification, it is apparent that he viewed animals as arranged from the simplest to the most complex in an ascending series. Since man was even more complex than animals and, moreover, possessed a rational faculty, he therefore occupied the highest position and a special category. This hierarchical perception of the animate world proved to be useful in every century to the present, except that in the modern view there is no such "scale of nature," and there is change in time by evolution from the simple to the complex.

After the time of Aristotle, Mediterranean science was centred at Alexandria, where the study of anatomy, particularly the central nervous system, flourished and, in fact, first became recognized as a discipline. Galen studied anatomy at Alexandria in the 2nd century and later dissected many animals. Much later, the contributions of the Renaissance anatomist Andreas Vesalius, though made in the context of medicine, as were those of Galen, stimulated to a great extent the rise of comparative anatomy. During the latter part of the 15th century and throughout the 16th century, there was a strong tradition in anatomy; important similarities were observed in the anatomy of different animals, and many illustrated books were published to record these observations.

But anatomy remained a purely descriptive science until the advent of functional considerations in which the correlation between structure and function was consciously investigated; as by French biologists Buffon and Cuvier. Cuvier cogently argued that a trained naturalist could deduce from one suitably chosen part of an animal's body the complete set of adaptations that characterized the organism. Because it was obvious that organisms with similar parts pursue similar habits, they were placed together in a system of classification. Cuvier pursued this viewpoint, which he called the theory of correlations, in a somewhat dogmatic manner and placed himself in opposition to the romantic natural philosophers, such as the German intellectual Johann Wolfgang von Goethe, who saw a tendency to ideal types in animal form. The tension between these schools of thought—adaptation as the consequence of necessary bodily functions and adaptation as an expression of a perfecting principle in nature—runs as a leitmotiv through much of biology, with overtones extending into the early 20th century.

The twin concepts of homology (similarity of origin) and analogy (similarity of appearance), in relation to structure, are the creation of the 19th-century British anatomist Richard Owen. Although they antedate the Darwinian view of evolution, the anatomical data on which they were based became, largely as a result of the work of the German comparative anatomist Carl Gegenbaur, important evidence in favour of evolutionary change, despite Owen's steady unwillingness to accept the view of diversification of life from a common origin.

Anatomy moved from a purely descriptive phase as an adjunct to classificatory studies, into a partnership with studies of function and became, in the 19th century, a major contributor to the concept of evolution.

Taxonomy or Systematics

Not until the work of Carolus Linnaeus did the variety of life receive a widely accepted systematic treatment. Linnaeus strove for a "natural method of arrangement," one that is now recognizable as an intuitive grasp of homologous relationships, reflecting evolutionary descent from a common ancestor; however, the natural method of arrangement sought by Linnaeus was more akin to the tenets of idealized morphology because he wanted to define a "type" form as epitomizing a species.

It was in the nomenclatorial aspect of classification that Linnaeus created a revolutionary advance with the introduction of a Latin binomial system: each species received a Latin name, which was not influenced by local names and which invoked the authority of Latin as a language common to the learned people of that day. The Latin name has two parts. The first word in the Latin name for the common chimpanzee, Pan troglodytes, for example, indicates the larger category, or genus, to which chimpanzees belong; the second word is the name of the species within the genus. In addition to species and genera, Linnaeus also recognized other classificatory groups, or taxa (singular taxon), which are still used; namely, order, class, and kingdom, to which have been added family (between genus and order) and phylum (between class and kingdom). Each of these can be divided further by the appropriate prefix of sub- or super-, as in subfamily or superclass. Linnaeus' great work, the Systema naturae, went through 12 editions during his lifetime; the 13th, and final, edition appeared posthumously. Although his treatment of the diversity of living things has been expanded in detail, revised in terms of taxonomic categories, and corrected in the light of continuing work—for example, Linnaeus treated whales as fish—it still sets the style and method, even to the use of Latin names, for contemporary nomenclatorial work.

Linnaeus sought a natural method of arrangement, but he actually defined types of species on the basis of idealized morphology. The greatest change from Linnaeus' outlook is reflected in the phrase "the new systematics," which was introduced in the 20th century and through which an explicit effort is made to have taxonomic schemes reflect evolutionary history. The basic unit of classification, the species, is also the basic unit of evolution—i.e., a population of actually or potentially interbreeding individuals. Such a population shares, through interbreeding, its genetic resources. In so doing, it creates the gene pool—its total genetic material—that determines the biological resources of the species and on which natural selection continuously acts. This approach has guided work on classifying animals away from somewhat arbitrary categorization of new species to that of recreating evolutionary history (phylogeny) and incorporating it in the system of classification. Modern taxonomists or systematists, therefore, are among the foremost students of evolution.

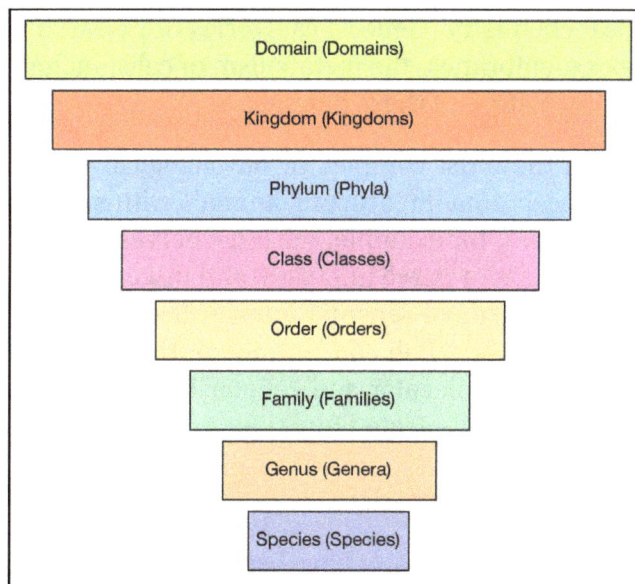

| Domain (Domains) |
| Kingdom (Kingdoms) |
| Phylum (Phyla) |
| Class (Classes) |
| Order (Orders) |
| Family (Families) |
| Genus (Genera) |
| Species (Species) |

Physiology

The practical consequences of physiology have always been an unavoidable human concern, in both medicine and animal husbandry. Inevitably, from Hippocrates to the present, practical knowledge of human bodily function has accumulated along with that of domestic animals and plants. This knowledge has been expanded, especially since the early 1800s, by experimental work on animals in general, a study known as comparative physiology. The experimental dimension had wide applications following Harvey's demonstration of the circulation of blood. From then on, medical physiology developed rapidly; notable texts appeared, such as Albrecht von Haller's eight-volume work Elementa Physiologiae Corporis Humani (Elements of Human Physiology), which had a medical emphasis. Toward the end of the 18th century the influence of chemistry on physiology became pronounced through Antoine Lavoisier's brilliant analysis of respiration as a form of combustion. This French chemist not only determined that oxygen was consumed by living systems but also opened the way to further inquiry into the energetics of living systems. His studies further strengthened the mechanistic view, which holds that the same natural laws govern both the inanimate and the animate realms.

Physiological principles achieved new levels of sophistication and comprehensiveness with Bernard's concept of constancy of the internal environment, the point being that only under certain constantly maintained conditions is there optimal bodily function. His rational and incisive insights were augmented by concurrent developments in Germany, where Johannes Müller explored the comparative aspects of animal function and anatomy, and Justus von Liebig and Carl Ludwig applied chemical and physical methods, respectively, to the solution of physiological problems. As a result, many useful techniques were advanced—e.g., means for precise measurement of muscular action and changes in blood pressure and means for defining the nature of body fluids.

By this time the organ systems—circulatory, digestive, endocrine, excretory, integumentary, muscular, nervous, reproductive, respiratory, and skeletal—had been defined, both anatomically and functionally, and research efforts were focussed on understanding these systems in cellular and chemical terms, an emphasis that continues to the present and has resulted in specialties in cell physiology and physiological chemistry. General categories of research now deal with the transportation of materials across membranes; the metabolism of cells, including synthesis and breakdown of molecules; and the regulation of these processes.

Interest has also increased in the most complex of physiological systems, the nervous system. Much comparative work has been done by utilizing animals with structures especially amenable to various experimental techniques; for example, the large nerves in squids have been extensively studied in terms of the transmission of nerve impulses, and insect and crustacean eyes have yielded significant information on patterns of sensory inputs. Most of this work is closely associated with studies on animal orientation and Behavior. Although the contemporary physiologist often studies functional problems at the molecular and cellular levels, he is also aware of the need to integrate cellular studies into the many-faceted functions of the total organism.

Embryology or Developmental Studies

Embryonic growth and differentiation of parts have been major biological problems since ancient times. A 17th-century explanation of development assumed that the adult existed as a miniature—a homunculus—in the microscopic material that initiates the embryo. But in 1759 the German physician Caspar Friedrick Wolff firmly introduced into biology the interpretation that undifferentiated materials gradually become specialized, in an orderly way, into adult structures. Although this epigenetic process is now accepted as characterizing the general nature of development in both plants and animals, many questions remain to be solved. The French physician Marie François Xavier Bichat declared in 1801 that differentiating parts consist of various components called tissues; with the subsequent statement of the cell theory, tissues were resolved into their cellular constituents. The idea of epigenetic change and the identification of structural components made possible a new interpretation of differentiation. It was demonstrated that the egg gives rise to three essential germ layers out of which specialized organs, with their tissues, subsequently emerge. Then, following his own discovery of the mammalian ovum, von Baer in 1828 usefully applied this information when he surveyed the development of various members of the vertebrate groups. At this point, embryology, as it is now recognized, emerged as a distinct subject.

The concept of cellular organization had an effect on embryology that continues to the present day. In the 19th century, cellular mechanisms were considered essentially to be the basis for growth, differentiation, and morphogenesis, or molding of parts. The distribution of the newly formed cells of the rapidly dividing zygote (fertilized egg) was precisely followed to provide detailed accounts not only of the time and mode of germ layer formation but also of the contribution of these layers to the differentiation of tissues and organs. Such descriptive information provided the background for experimental work aimed at elucidating the role of chromosomes and other cellular constituents in differentiation. About 1895, before the formulation of the chromosomal theory of heredity, Theodor Boveri demonstrated that chromosomes show continuity from one cell generation to the next. In fact, biologists soon concluded that in all cells arising from a fertilized egg, half the chromosomes are of maternal and half of paternal origin. The discovery of the constant transmission of the original chromosomal endowment to all cells of the body served to deepen the mystery surrounding the factors that determine cellular differentiation.

The present view is that differential activity of genes is the basis for cellular and tissue differentiation; that is, although the cells of a multicellular body contain the same genetic information, different genes are active in different cells. The result is the formation of various gene products, which regulate the functional and structural differentiation of cells. The actual mechanism involved in the inactivation of certain genes and the activation of others, however, has not yet been established. That cells can move extensively throughout the embryo and selectively adhere to other cells, thus starting tissue aggregations, also contributes to development as does the fate of cells—i.e., certain ones continue to multiply, others stop, and some die.

Research methods in embryology now exploit many experimental situations: both unicellular and multicellular forms; regeneration (replacement of lost parts) and normal development; and growth of tissues outside and inside the host. Hence, the processes of development can be studied with material other than embryos; and the study of embryology has become incorporated into the more inclusive subdiscipline of developmental biology.

Evolutionism

Darwin was not the first to speculate that organisms can change from generation to generation and so evolve, but he was the first to propose a mechanism by which the changes are accumulated. He proposed that heritable variations occur in conjunction with a never-ending competition for survival and that the variations favouring survival are automatically preserved. In time, therefore, the continued accumulation of variations results in the emergence of new forms. Because the variations that are preserved relate to survival, the survivors are highly adapted to their environment. To this process Darwin gave the apt name natural selection.

Many of Darwin's predecessors, notably Jean-Baptiste Lamarck, were willing to accept the idea of species variation, even though to do so meant denying the doctrine of special creation and the static-type species of Linnaeus. But they argued that some idealized perfecting principle, expressed through the habits of an organism, was the basis of variation. The contrast between the romanticism of Lamarck and the objective analysis of Darwin clearly reveals the type of revolution provoked by the concept of natural selection. Although mechanistic explanations had long been available to biologists—forming, for example, part of Harvey's explanation of blood circulation—they did not pervade the total structure of biological thinking until the advent of Darwinism.

There were two immediate consequences of Darwin's viewpoints. One has involved a reappraisal of all subject areas of biology; reinterpretations of morphology and embryology are good examples. The comparative anatomy of the British anatomist Owen became a cornerstone of the evidence for evolution, and German anatomists provided the basis for the comment that evolutionary thinking was born in England but gained its home in Germany. The reinterpretation of morphology carried over into the study of fossil forms, as paleontologists sought and found evidence of gradual change in their study of fossils. But some workers, although accepting evolution in principle, could not easily interpret the changes in terms of natural selection. The German paleontologist Otto Schindewolf, for example, found in shelled mollusks called ammonites evidence of progressive complexity and subsequent simplification of forms. The American paleontologist George Gaylord Simpson, however, has been a consistent interpreter of vertebrate fossils by Darwinian selection. Embryology was seen in an evolutionary light when the German zoologist Ernst Haeckel proposed that the epigenetic sequence of embryonic development (ontogeny) repeated its evolutionary history (phylogeny). Thus, the presence of gill clefts in the mammalian embryo and also in less highly evolved vertebrates can be understood as a remnant of a common ancestor.

The other consequence of Darwinism—to make more explicit the origin and nature of heritable variations and the action of natural selection on them—depended on the emergence of the following: genetics and the elucidation of the rules of Mendelian inheritance; the concept of the gene as the unit of inheritance; and the nature of gene mutation. The development of these ideas provided the basis for the genetics of natural populations.

The subject of population genetics began with the Mendelian laws of inheritance and now takes into account selection, mutation, migration (movement into and out of a given population), breeding patterns, and population size. These factors affect the genetic makeup of a group of organisms that either interbreed or have the potential to do so i.e., a species. Accurate appraisal of these factors allows precise predictions regarding the content of a given gene pool over significant periods of evolutionary time. From work involving population genetics has come the realization, eloquently documented by two contemporary American evolutionists, Theodosius Dobzhansky and Ernst Mayer, that the species is the basic unit of evolution. The process of speciation occurs as a gene pool breaks up to form isolated gene pools. When selection pressures similar to those of the original gene pool persist in the new gene pools, similar functions and the similar structures on which they depend also persist. When selection pressures differ, however, differences arise. Thus, the process of speciation through natural selection preserves the evolutionary history of a species. The record may be discerned not only in the gross, or macroscopic, anatomy of organisms but also in their cellular structure and molecular organization. Significant work now is carried out, for example, on the homologies of the nucleic acids and proteins of different species.

Genetics

The problem of heredity had been the subject of careful study before its definitive analysis by Mendel.

As with Darwin's predecessors, those of Mendel tended to idealize and interpret all inherited traits as being transmitted through the blood or as determined by various "humors" or other vague entities in animal organisms. When studying plants, Mendel was able to free himself of anthropomorphic and holistic explanations. By studying seven carefully defined pairs of characteristics—e.g., tall and short plants; red and white flowers, etc.—as they were transmitted through as many as three successive generations, he was able to establish patterns of inheritance that apply to all sexually reproducing forms. Darwin, who was searching for an explanation of inheritance, apparently never saw Mendel's work, which was published in 1866 in the obscure journal of his local natural history society; it was simultaneously rediscovered in 1900 by three different European geneticists.

Further progress in genetics was made early in the 20th century, when it was realized that heredity factors are found on chromosomes. The term gene was coined for these factors. Studies by the American geneticist Thomas Hunt Morgan on the fruit fly (Drosophila), moved animal genetics to the forefront of genetic research. The work of Morgan and his students established such major concepts as the linear array of genes on chromosomes; the exchange of parts between chromosomes; and the interaction of genes in determining traits, including sexual differences. In 1927 one of Morgan's former students, Hermann Muller, used X rays to induce the mutations (changes in genes) in the fruit fly, thereby opening the door to major studies on the nature of variation.

Meanwhile, other organisms were being used for genetic studies, most notably fungi and bacteria. The results of this work provided insights into animal genetics just as principles initially obtained from animal genetics provided insight into botanical and microbial forms. Work continues not only on the genetics of humans, domestic animals, and plants but also on the control of development through the orderly regulation of gene action in different cells and tissues.

Cellular and Molecular Biology

Although the cell was recognized as the basic unit of life early in the 19th century, its most exciting period of inquiry has probably occurred since the 1940s. The new techniques developed since that time, notably the perfection of the electron microscope and the tools of biochemistry, have changed the cytological studies of the 19th and early 20th centuries from a largely descriptive inquiry, dependent on the light microscope, into a dynamic, molecularly oriented inquiry into fundamental life processes.

The so-called cell theory, which was enunciated about 1838, was never actually a theory. As Edmund Beecher Wilson, the noted American cytologist, stated in his great work, The Cell,

> "By force of habit we still continue to speak of the cell 'theory' but it is a theory only in name. In substance it is a comprehensive general statement of fact and as such stands today beside the evolution theory among the foundationstones of modern biology."

More precisely, the cell doctrine was an inductive generalization based on the microscopial examination of certain plant and animal species.

Rudolf Virchow, a German medical officer specializing in cellular pathology, first expressed the fundamental dictum regarding cells in his phrase omnis cellula e cellula (all cells from cells). For cellular reproduction is the ultimate basis of the continuity of life; the cell is not only the basic structural unit of life but also the basic physiological and reproductive unit. All areas of biology were affected by the new perspective afforded by the principle of cellular organization. Especially in conjunction with embryology was the study of the cell most prominent in animal biology. The continuity of cellular generations by reproduction also had implications for genetics.

The study of the cell nucleus, its chromosomes, and their Behavior served as the basis for understanding the regular distribution of genetic material during both sexual and asexual reproduction. This orderly Behavior of the nucleus made it appear to dominate the life of the cell, for by contrast the components of the rest of the cell appeared to be randomly distributed.

The biochemical study of life had helped in the characterization of the major molecules of living systems—proteins, nucleic acids, fats, and carbohydrates—and in the understanding of metabolic processes. That nucleic acids are a distinctive feature of the nucleus was recognized after their discovery by the Swiss biochemist Johann Friedrich Miescher in 1869. In 1944 a group of American bacteriologists, led by Oswald T. Avery, published work on the causative agent of pneumonia in mice (a bacterium) that culminated in the demonstration that deoxyribonucleic acid (DNA) is the chemical basis of heredity. Discrete segments of DNA correspond to genes, or Mendel's hereditary factors. Proteins were discovered to be especially important for their role in determining cell structure and in controlling chemical reactions.

The advent of techniques for isolating and characterizing proteins and nucleic acids now allows a molecular approach to essentially all biological problems—from the appearance of new gene products in normal development or under pathological conditions to a monitoring of changes in and between nerve cells during the transmission of nerve impulses.

Ecology

The harmony that Linnaeus found in nature, which redounded to the glory and wisdom of a Judaeo-Christian god, was the 18th-century counterpart of the balanced interaction now studied by ecologists. Linnaeus recognized that plants are adapted to the regions in which they grow, that insects play a role in flower pollination, and that certain birds prey on insects and are in turn eaten by other birds. This realization implies, in contemporary terms, the flow of matter and energy in a definable direction through any natural assemblage of plants, animals, and microorganisms. Such an assemblage, termed an ecosystem, starts with the plants, which are designated as producers because they maintain and reproduce themselves at the expense of energy from sunlight and inorganic materials taken from the nonliving environment around them (earth, air, and water). Animals are called consumers because they ingest plant material or other animals that feed on plants, using the energy stored in this food to sustain themselves. Lastly, the organisms known as decomposers, mostly fungi and bacteria, break down plant and animal material and return it to the environment in a form that can be used again by plants in a constantly renewed cycle.

The term ecology, first formulated by Haeckel in the latter part of the 19th century as "oecology" referred to the dwelling place of organisms in nature. In the 1890s various European and U.S. scientists laid the foundations for modern work through studies of natural ecosystems and the populations of organisms contained within them.

Animal ecology, the study of consumers and their interactions with the environment, is very complex; attempts to study it usually focus on one particular aspect. Some studies, for example, involve the challenge of the environment to individuals with special adaptations (e.g., water conservation in desert animals); others may involve the role of one species in its ecosystem or the ecosystem itself. Food-chain sequences have been determined for various ecosystems, and the efficiency of the transfer of energy and matter within them has been calculated so that their capacity is known; that is, productivity in terms of numbers of organisms or weight of living matter at a specific level in the food chain can be accurately determined.

In spite of advances in understanding animal ecology, this subject area of zoology does not yet have the major unifying theoretical principles found in genetics (gene theory) or evolution (natural selection).

Ethology

The study of animal Behavior (ethology) is largely a 20th-century phenomenon and is exclusively a zoological discipline. Only animals have nervous systems, with their implications for perception, coordination, orientation, learning, and memory. Not until the end of the 19th century did animal Behavior become free from anthropocentric interests and assume an importance in its own right. The British behaviorist C. Lloyd Morgan was probably most influential with his emphasis on parsimonious explanations—i.e., that the explanation "which stands lower in the psychological scale" must be invoked first. This principle is exemplified in the American Herbert Spencer Jennings' pioneering work in 1906 on- The Behavior of Lower Organisms.

The study of animal Behavior now includes many diverse topics, ranging from swimming patterns of protozoans to socialization and communication among the great apes. Many disparate hypotheses have been proposed in an attempt to explain the variety of behavioral patterns found in animals. They focus on the mechanisms that stimulate courtship in reproductive Behavior of such diverse groups as spiders, crabs, and domestic fowl; and on whole life histories, starting from the special attachment of newly born ducks and goats to their actual mothers or to surrogate (substitute) mothers. The latter phenomenon, called imprinting, has been intensively studied by the Austrian ethologist Konrad Lorenz. Physiologically oriented Behavior now receives much attention; studies range from work on conditioned reflexes to the orientation of crustaceans and the location and communication of food among bees; such diversity of material is one measure of the somewhat diffuse but exciting current state of these studies.

Zoology has become animal biology—that is, the life sciences display a new unity, one that is founded on the common basis of all life, on the gene pool–species organization of organisms, and on the obligatory interacting of the components of ecosystems. Even as regards the specialized features of animals—involving physiology, development, or Behavior—the current emphasis is on elucidating the broad biological principles that identify animals as one aspect of nature. Zoology has thus given up its exclusive emphasis on animals—an emphasis maintained from Aristotle's time well into the 19th century—in favour of a broader view of life. The successes in applying physical and chemical ideas and techniques to life processes have not only unified the life sciences but have also created bridges to other sciences in a way only dimly foreseen by earlier workers. The practical and theoretical consequences of this trend have just begun to be realized.

Methods in Zoology

Because the study of animals may be concentrated on widely different topics, such as ecosystems and their constituent populations, organisms, cells, and chemical reactions, specific techniques are needed for each kind of investigation. The emphasis on the molecular basis of genetics, development, physiology, Behavior, and ecology has placed increasing importance on those techniques involving cells and their many components. Microscopy, therefore, is a necessary technique in zoology, as are certain physicochemical methods for isolating and characterizing molecules. Computer technology also has a special role in the analysis of animal life. These newer techniques are used in addition to the many classical ones—measurement and experimentation at the tissue, organ, organ system, and organismic levels.

Microscopy

In addition to continuous improvements in the techniques of staining cells, so that their components can be seen clearly, the light used in microscopy can now be manipulated to make visible certain structures in living cells that are otherwise undetectable. The ability to observe living cells is an advantage of light microscopes over electron microscopes; the latter require the cells to be in an environment that kills them. The particular advantage of the electron microscope, however, is its great powers of magnification. Theoretically, it can resolve single atoms; in biology, however, magnifications of lesser magnitude are most useful in determining the nature of structures lying between whole cells and their constituent molecules.

Separation and Purification Techniques

The characterization of components of cellular systems is necessary for biochemical studies. The specific molecular composition of cellular organelles, for example, affects their shape and density (mass per unit volume); as a result, cellular components settle at different rates (and thus can be separated) when they are spun in a centrifuge.

Other methods of purification rely on other physical properties. Molecules vary in their affinity for the positive or negative pole of an electrical field. Migration to or away from these poles, therefore, occurs at different rates for different molecules and allows their separation; the process is called electrophoresis. The separation of molecules by liquid solvents exploits the fact that the molecules differ in their solubility, and hence they migrate to various degrees as a solvent flows past them. This process, known as chromatography because of the colour used to identify the position of the migrating materials, yields samples of extraordinarily high purity.

Radioactive Tracers

Radioactive compounds are especially useful in biochemical studies involving metabolic pathways of synthesis and degradation. Radioactive compounds are incorporated into cells in the same way as their nonradioactive counterparts. These compounds provide information on the sites of specific metabolic activities within cells and insights into the fates of these compounds in both organisms and the ecosystem.

Computers

Computers process information using their own general language, which is able to complete calculations as complex and diverse as statistical analyses and determinations of enzymatically controlled reaction rates. Computers with access to extensive data files can select information associated with a specific problem and display it to aid the researcher in formulating possible solutions. They help perform routine examinations such as scanning chromosome preparations in order to identify abnormalities in number or shape. Test organisms can be electronically monitored with computers, so that adjustments can be made during experiments; this procedure improves the quality of the data and allows experimental situations to be fully exploited. Computer simulation is important in analyzing complex problems; as many as 100 variables, for example, are involved in the management of salmon fisheries. Simulation makes possible the development of models that approach the complexities of conditions in nature, a procedure of great value in studying wildlife management and related ecological problems.

Applied Zoology

Animal-related industries produce food (meats and dairy products), hides, furs, wool, organic fertilizers, and miscellaneous chemical byproducts. There has been a dramatic increase in the productivity of animal husbandry since the 1870s, largely as a consequence of selective breeding and improved animal nutrition. The purpose of selective breeding is to develop livestock whose desirable traits have strong heritable components and can therefore be propagated. Heritable components are distinguished from environmental factors by determining the coefficient of heritability, which is defined as the ratio of variance in a gene-controlled character to total variance.

Another aspect of food production is the control of pests. The serious side effects of some chemical pesticides make extremely important the development of effective and safe control mechanisms. Animal food resources include commercial fishing. The development of shellfish resources and fisheries management (e.g., growth of fish in rice paddies in Asia) are important aspects of this industry.

2

Branches of Zoology

Herpetology, mammalogy, ornithology, entomology, ichthyology, primatology and anthrozoology are some of the branches of zoology. All these branches of zoology have been thoroughly discussed in this chapter to give a better understanding of the subject.

MAMMALOGY

Mammalogy is the branch of biology that deals with the study of mammals. It encompasses such diverse areas as the structure, function, evolutionary history, ethology, taxonomy, management and economics of mammals. Approximately 4,200 species of living mammals and numerous extinct species comprise the material for study. Included are egg-laying echidnas, the platypus, pouched marsupials, tiny shrews, bats, mice, whales, apes and elephants, to name only a few.

The study of mammals can be as diverse as the organisms themselves. A mammalogist might study a wide variety of topics on a particular species or group of mammals or might take a comparative approach and investigate one aspect with regard to a wide variety of mammals. The major subdivisions of the science of mammalogy include the following:

Taxonomy and Systematics

The study of the classification of mammals into distinct orders, families, genera, species and subspecies, thereby defining the geographic distribution of each taxon (a taxonomic category or group, such as a phylum, order, family, genus or species). In addition, the evolutionary relationships among extinct (paleontology) and/or living taxa are analyzed.

Anatomy and Physiology

The study of mammalian body structures and tissues and how each functions.

Ethology

The study of the behavior of an animal and how those behaviors influence its survival and reproduction.

Ecology

The study of interactions between mammals and their environments (nonliving and living). This discipline includes the study of special adaptations to environmental factors (physiological ecology) and the study of interactions within and among species (population and community ecology).

Management and Control

The interactions between humans and other mammals in which humans manipulate either the environment or populations of mammals to favor the use and/or survival of certain species and to regulate or even reduce the populations of species whose activities conflict with human interests.

PRIMATOLOGY

Primatology is the study of the primate order of mammals—other than recent humans (Homo sapiens). The species are characterized especially by advanced development of binocular vision, specialization of the appendages for grasping, and enlargement of the cerebral hemispheres.

Nonhuman primates provide a broad comparative framework within which physical anthropologists can study aspects of the human career and condition. Comparative morphological studies, particularly those that are complemented by biomechanical analyses, provide major clues to the functional significance and evolution of the skeletal and muscular complexes that underpin humans' bipedalism, dextrous hands, bulbous heads, outstanding noses, and puny jaws. The wide variety of adaptations that primates have made to life in trees and on the ground are reflected in their limb proportions and relative development of muscles.

Free-ranging primates exhibit a trove of physical and behavioral adaptations to fundamentally different ways of life, some of which may resemble those of humans' late Miocene–early Pleistocene predecessors (i.e., those from about 11 to 2 million years ago). Laboratory and field observations, particularly of great apes, indicate that earlier researchers grossly underestimated the intelligence, cognitive abilities, and sensibilities of nonhuman primates and perhaps also those of Pliocene–early Pleistocene hominins (i.e., those from about 5.3 to 2 million years ago), who left few archaeological clues to their Behavior.

ANTHROZOOLOGY

Anthrozoology is the study of the interactions and relationships between human and nonhuman animals. Anthrozoology spans the humanities and the social, behavioral, and biomedical sciences.

While the lives of humans and nonhuman animals have always been intertwined, the ways that humans relate to and think about members of other species became the focus of systematic study only in the late 20th century. The development of anthrozoology as an academic discipline was spurred by reports that there were health and psychological benefits to interacting with animals and by the establishment of the academic journals Anthrozoös in 1987 and Society & Animals in 1993.

Topics of anthrozoological inquiry include the psychological and biological underpinnings of attachments to pets, attitudes toward the use of animals, cross-cultural similarities and differences in human-animal relationships, sex differences in interactions with other species, and the roles of animals in art, religion, mythology, sport, and literature. The impact of companion animals on human health and happiness continues to be an active area of research. Interacting with pets has been found to lower their owners' blood pressure and stress levels, and pet ownership is associated with increased survivorship following heart attacks. Some epidemiological studies have reported that pet owners make fewer visits to doctors, are more physically active, and have lower levels of depression. However, research on the effects of pets on humans has produced conflicting results, and some studies have found that the health and happiness of pet owners was no better, and in some cases was worse, than non-pet owners'.

Interacting with animals has been used as a therapeutic intervention for disorders such as autism and to enhance morale in institutions such as nursing homes, hospitals, and prisons. Relatively few randomized clinical trials (in which participants are assigned at random to different treatments) have assessed the long-term effectiveness of animal-assisted therapy, but it is clear that many people find stroking, talking to, and playing with companion animals deeply satisfying.

The link between animal abuse and aggression in humans is also an important area of research. Some investigators have reported that childhood animal cruelty is a strong predictor of violent Behavior in adults. Other researchers, however, argue that most children who abuse animals do not become violent adults and that the relationship between animal cruelty and human-directed aggression is not as strong as was originally thought.

An autistic teenager holds his puppy and a small bucket of his belongings. Interaction with animals is sometimes used as a therapeutic intervention for persons with disorders such as autism.

ICHTHYOLOGY

Ichthyology is the branch of zoology devoted to the study of fish. This includes bony fish (class Osteichthyes, with over 26,000 species), cartilaginous fish (class Chondrichthyes, about 800 species including sharks and rays), and jawless fish (class or superclass Agnatha, about 75 species including lampreys and hagfish).

The study of fish, which is centuries old, reveals humanity's strong and lasting curiosity about nature, with fish providing both inner joy (beauty, recreation, wonder, and religious symbolism) and practical values (ecology, food, and commerce).

With about 27,000 known living species, fish are the most diverse group of vertebrates, with more than one-half of the total vertebrate species. While a majority of species have probably been discovered and described, approximately 250 new species are officially described by science each year.

Hagfish, while generally classified in Agnatha and as fish, actually lack vertebrae, and for this reason sometimes are not considered to be fish. Nonetheless, they remain a focus of ichthyology. Many types of aquatic animals named "fish," such as jellyfish, starfish, and cuttlefish, are not true fish. They, and marine mammals like whales, dolphins, and pinnipeds (seals and walruses) are not a focus of ichthyology.

The practice of ichthyology is associated with aquatic biology, limnology, oceanography, and aquaculture.

Ichthyology originated near the beginning of the Upper Paleolithic period, about forty thousand years ago, and continues to the present day. This science was developed in several interconnecting epochs, each with various significant advancements. According to K.F. Lagler et al., the study of fishes (ichthyology) was hardly scientific until the eighteenth century. However, there were attempts to study fish, if only to learn how to propagate them for aquaculture, to capture them by fishing, or to adhere to dietary laws.

Early Developments

The study of fish likely receives its origins from the human desire to feed, clothe, and equip them-selves with useful implements. Early ichthyologists were likely hunters and gatherers who inves-tigated which fish were edible, where they could be found, and how to best capture them. These insights of early cultures were manifested in abstract and identifiable artistic expressions.

Around 3,500 B.C.E. the Chinese were trying to learn about fish in order to practice aquaculture. When the waters lowered after river floods, some fishes, namely carp, were held in artificial lakes. Their brood were later fed using nymphs and feces from silkworms used for silk production.

There is evidence of Egyptian aquaculture, focusing on tilapia, tracing to 2000 B.C.E.

Moses, in the development of the kashrut (Jewish dietary laws), forbade the consumption of fish without scales or appendages. This required some study of fish that has continued to this day.

Fish compose approximately 8% of all figurative depictions on Mimbres pottery.

Foundation of Formal Study

The oldest known document on fish culture was written by a Chinese politician, Fan-Li, in 475 B.C.E. Aristotle incorporated ichthyology into formal scientific study. Between 335 B.C.E. and 322 B.C.E., he provided the earliest taxonomic classification of fish, in which 117 species of Mediterranean fish were accurately described. Furthermore, Aristotle observed the anatomical and behavioral differences between fish and marine mammals.

After his death, some of Aristotle's pupils continued his ichthyological research. Theophrastus, for example, composed a treatise on amphibious fishes.

The Romans, although less devoted to the pursuit of science than the Greeks, wrote extensively about fish. Pliny the Elder, a notable Roman naturalist, compiled the ichthyological works of indigenous Greeks, including verifiable and ambiguous peculiarities such as the sawfish and mermaid, respectively. During this time, the study of fish was also pursued in less systematic ways, either for fishing or aquaculture.

Roman aquaculture was practiced in the first century B.C.E., according to Pliny the Elder. The Romans focused on trout and mullet and were quite adept at breeding fish in ponds.

Theologians and ichthyologists speculate that the apostle Peter and his contemporaries harvested the fish that are today sold in modern industry along the Sea of Galilee, presently known as Lake Kinneret. These fish include cyprinids of the genus Barbus and Mirogrex, cichlids of the genus Sarotherodon, and Mugil cephalus of the family Mugilidae.

The Hawaiian people practiced aquaculture by constructing fish ponds, with an organized system in place by 400 C.E. A remarkable example from ancient Hawaii is the construction of a fish pond, dating from at least 1,000 years ago, at Alekoko.

In Central Europe, there is record of pond fish culture at the end of the eleventh century C.E.

In Europe during the Middle Ages, aquaculture became common in monasteries, as fish was scarce and thus expensive. A fourteenth century French monk, Dom Pinchon, may have been the first person to artificially fertilize trout eggs. There is also evidence that the Maya had a form of aquaculture, as did the native peoples of North America.

The Development of Modern Ichthyology

The writings of three sixteenth century scholars, Hippolyte Salviani, Pierre Belon, and Guillaume Rondelet, signify the conception of modern ichthyology. The investigations of these individuals were based upon actual research in comparison to ancient recitations. Despite their prominence, Rondelet's De Piscibus Marinum is regarded as the most influential, identifying 244 species of fish.

The incremental alterations in navigation and shipbuilding throughout the Renaissance marked the commencement of a new epoch in ichthyology. The Renaissance culminated with the era of exploration and colonization, and upon the cosmopolitan interest in navigation came the specialization in naturalism.

Georg Marcgrave of Saxony composed the Naturalis Brasilae in 1648. This document contained a description of one hundred species of fish indigenous to the Brazilian coastline. In 1686 John Ray and Francis Willughby collaboratively published Historia Piscium, a scientific manuscript containing 420 species of fish, 178 of these newly discovered. The fish contained within this informative literature were arranged in a provisional system of classification.

The classification used within the Historia Piscium was improved upon by Carolus Linnaeus, the "father of modern taxonomy." His two prime contributions were: (1) to establish conventions for the naming of living organisms using binomial nomenclature (the genus name followed by the species name), and (2) developing a hierarchical system for classification of organisms. Although the system now known as binomial nomenclature was developed by the Bauhin brothers (Gaspard Bauhin and Johann Bauhin) almost two hundred years earlier, Linnaeus was the first to use it consistently, and may be said to have popularized it within the scientific community. Linnaeus's taxonomic approach became the systematic approach to the study of organisms, including fish.

It was one of Linnaeus's colleagues, Peter Artedi, who earned the title "father of ichthyology" through his indispensable advancements. Artedi contributed to Linnaeus's refinement of the principles of taxonomy. Furthermore, he recognized five additional orders of fish: Malacopterygii, Acanthopterygii, Branchiostegi, Chondropterygii, and Plagiuri. Artedi developed standard methods for making counts and measurements of anatomical features that are modernly exploited. Another associate of Linnaeus, Albertus Seba, was a prosperous pharmacist from Amsterdam. Seba assembled a cabinet, or collection, of fish. He invited Artedi to utilize this assortment of fish; unfortunately, in 1735, Artedi fell into an Amsterdam canal and drowned at the age of 30. Linnaeus posthumously published Artedi's manuscripts as Ichthyologia, sive Opera Omnia de Piscibus (1738).

Linnaeus revised the orders introduced by Artedi, placing significance on pelvic fins. Fish lacking this appendage were placed within the order Apodes; fish containing abdominal, thoracic, or jugular pelvic fins were termed Abdominales, Thoracici, and Jugulares respectively. However, these alterations were not grounded within the evolutionary theory. Therefore, it would take over a century until Charles Darwin would provide the intellectual foundation from which it would be perceived that the degree of similarity in taxonomic features corresponded to phylogenetic relationship.

Modern Era

Close to the dawn of the nineteenth century, Marcus Elieser Bloch of Berlin and Georges Cuvier of Paris made an attempt to consolidate the knowledge of ichthyology. Cuvier summarized all of the available information in his monumental Histoire Naturelle des Poissons. This manuscript was published between 1828 and 1849 in a 22 volume series. This documentation contained 4,514 species of fish, 2,311 of these new to science. This piece of literature still remained one of the most ambitious treatises of the modern world.

The scientific exploration of the Americas advanced knowledge of the remarkable diversity of fish. Charles Alexandre Lesueur, a student of Cuvier, made a collection of fish dwelling within the Great Lakes and Saint Lawrence River regions.

Adventurous individuals such as John James Audubon and Constantine Samuel Rafinesque figure into the faunal documentation of North America. These persons often traveled with one another and composed Ichthyologia Ohiensis in 1820. In addition, Louis Agassiz of Switzerland established his reputation through the study of freshwater fish and organisms and the pioneering of paleoichthyology. Agassiz eventually immigrated to the United States and taught at Harvard University in 1846.

Albert Günther published his Catalogue of the Fishes of the British Museum between 1859 and 1870, describing over 6,800 species and mentioning another 1,700. Generally considered one of the most influential ichthyologists, David Starr Jordan wrote 650 articles and books on the subject as well as serving as president of Indiana University and Stanford University.

Today, ichthyology is a well-known scientific field investigating such areas related to fish as classification, anatomy, evolution, genetics, ecology, physiology, and conservation.

ENTOMOLOGY

Entomology is a branch of zoology (the study of animals) that studies insects and how they interact with their environment, other species and humans. An animal is an insect when it has segmented body parts. Insects are quite remarkable creatures and we have identified in the region of 1 million species with estimates of the number of unidentified species ranging from 5 million to 8 million. Insects exist all over the world and survive in some of the harshest environments on earth; it is believed that they - in terms of numbers - outnumber all other animal species combined.

Humans have always been interested in insects for one reason or another; ancient cultures have examined, farmed and even venerated them. Ancient Egyptians worshipped a large species of dung beetle, or Scarab, that would gather balls of dung and bury them. The female would lay eggs on the dung, and then weeks or months later, new beetles would emerge from the ground - seemingly reborn from nothing and thus representing the renewal of life. Even before this, some of the earliest cave art depicts bees. Many Roman writers discussed insects and Aristotle and Pliny the Elder both had a strong fascination in which they published their observations of insects in books on natural history.

The true scientific study of insects did not development until the Renaissance and through to the Enlightenment era - with the greatest expansion of the subject seeming to be the 1800s. Three strands of study grew up in a very short space of time: first there were the researchers who wanted to depict nature's beauty through producing highly detailed image. This wasn't just useful for art's sake, these highly detailed sketches were useful to researchers wanting to understand the insects' physiology.

The second group concerned itself (as many natural sciences did) with classification, dividing for the sake of convenience and ease of study - still a common practice today and we still use the division method in classification of new species, even where and when we realise that not all species of anything fit neatly into boxes. The third group focused largely on examining the biological processes of insects - life cycle, reproduction, habitats and other elements.

This developed through the course of the 18th century and by the 19th century most major universities were studying the world's insects and major scientific institutes began a systematic programme of research and investigation. In North America and in Europe, the Victorian era saw the greatest expansion of interest of insects firstly through the amateur research of the well to do gentleman researcher, and later into more academic study leading to the formalised science that we have today.

There is no greater work with such a wide scope for this third group, than the four volumes published in the early 19th century by William Kirby and William Spence and it is still considered a seminal work today. For the most part, this 19th century research into insects was driven by medical science and a need to understand and combat diseases carried by insects such as malaria and yellow fever, their causes and how they spread. It was important to the colonialism and the growing market economy of nations spreading their influence and building markets in developing countries. That's not to say that study for study's sake was not an issue as it was with so many other 18th and 19th century naturalists, because it was. As humans we are driven to examining and understanding the world around us.

The 20th and 21st centuries led to the more discoveries of insect species, and how they live and reproduce, than ever before. Now we can visit very remote areas far more easily than we could before - new species turn up every day. We have also discovered that the study of insects has uses outside of entomology.

What is an Insect?

It may seem like a simple question, but there is some confusion over what is and what is not an insect. There are certain criteria by which we define which creatures are insects. For starters, they must have an exoskeleton - this is common to all arthropods. In order for the arthropod to be an insect, the specimen must have six legs - this is generally what separates them from other arthropods. They also have three distinct body parts broken down into the head, abdomen and thorax. They may or may not have wings, or antennae, or both and they may live on land, in the air or in bodies of water (though there are not many marine species as these environments are dominated by other types of arthropods).

Most insects have compound eyes that are large relative to their bodies, though some are eyeless and many have ocelli (sensors that fulfill some functions of eyes in other species). Combined, the compound eyes, antennae and ocelli perform most sensory functions of the insect. Sensory hairs on the bodies of many insects tell them the direction the wind is blowing, so if they smell food, they know what direction to fly or crawl to find it.

"Not all insects have ocelli - in fact, this is a distinguishing character at the taxonomic level of orders. Some type of beetles for example such as the subfamily Omaliinae of the Staphylinidae do have ocelli while most beetle types do not. Beetles about 40% of all named insect species."

What can Entomology Tell us?

The subject does not exist in a bubble where it is useful only to itself; it is part of a wider environmental study of our natural ecology and human geography - insects are affected by the climate and results of studies can tell us much about what is going on in the world around us. We discover new uses for studies in entomology all the time and there are already several established disciplines where insects and insect remains are a powerful type of evidence.

As Environmental Indicators

Insects are a vital part of forest biodiversity and as they are particularly sensitive to changes in the climate and the patterns and seasons of regional flora cover, entomology provides evidence for general forest health as well changes in the cover. The arrival of a new pest or the sudden decline of a well-established native species can indicate many things, such as the effects of deforestation, or a change in the types of arboreal cover, CO_2 concentration, persistent ecological issues such as drought or flooding. They can tell us much about the types of trees, shrubs and flowers that grow there (especially pollinating insects such as butterflies) and any changes.

The Arctic Circle has a very delicate ecosystem and this is one area where researchers in many fields rely heavily on entomology, particularly of the relationship between insect groups, between insects and the environment, between insects and other animal species; evidence from population numbers and density is often used to examine specific environmental indicators.

For Archaeology/Anthropology

Similarly, and as archaeology moves more into the realm of environmental science, insect population and remains are becoming more important in the interpretation of archaeological sites and landscapes. Humans have had an enormous impact on the environment of the past, and we are just beginning to understand how geoengineering, how hunter-gatherer societies and even the Neolithic Agricultural Revolution impacted local ecologies. We are only now beginning to understand how insect remains are useful, and indeed which insect species are most useful in which areas. In a study of western Greenland for example, researchers examined extant insect remains for indicators of personal hygiene.

This isn't all they can tell us though. When a virgin landscape is altered to make way for farmland, there is likely to be a change to the ecosystem that will include a change of both the pollen distribution pattern and insect remains - especially where a crop favourable to native insect species replaces a crop that was unfavourable, or vice versa. In these cases, there will be a clear and distinct pattern change in the archaeological record.

Forensic Entomology

It may surprise people to learn that insect species can sometimes play a part in criminal investigation and even be included as part of the evidence. We can learn much about a body's season of death and for how long it may have been exposed to the elements before discovery or being buried. This is one of the most critical areas in which entomology contributes to our understanding of human remains - in archaeology and in criminal proceedings.

Insects are attracted to decomposing bodies and may begin feeding off it or laying eggs in it. By understanding the lifecycle process of any insect remains found, investigators can identify how long a body has been dead, where it has been storied and a variety of other facts. Most interestingly, it seems that studying insects as a method of working out the perpetrator of a murder goes back to China in the 13th century when suspects were told to lay their sickles on the ground to see which attracted the most flies to the small amounts of blood that remained on the blade. Using insects as evidence in a modern criminal trial began in 1935 when a body in the UK showed a number of blowflies. The subsequent evidence attained from the blowflies informed investigators how long the woman had been dead and led to a conviction of the guilty party.

Medical Entomology

This is the area concerned with medical research and how insects are a matter of public health. This area also includes plant disease and damage caused by insects to crops too. That means researchers in Medical Entomology are concerned with urban and rural pests as well as the diseases that some insects carry. The scope of medical entomology will also cover research into the effectiveness of, and development of new, pesticides.

Medical entomology arguably began in the early 20th century when it was discovered that mosquitoes were responsible for the transmission of Yellow Fever. Medical entomology plays a large part of the US military research as the major bodies examined and combated the effects of major diseases in all wars the country has participated in - most recently in Iraq and Afghanistan. Though entomology is the study of insects, medical entomology has a broader scope in that it incorporates other arthropods that may affect human health - this means arachnids such as spiders, mites, ticks and also come under the scope of a medical entomology researcher.

Fruit Fly Research

The humble fruit fly deserves a special mention because these small insects have given us so much information. We have learnt much about evolutionary biology because many generations breed in a very short space of time. We have information on genetic mutation, genetic drift and other elements of evolution vital to research in other areas, particularly human genetics. Selective cross-breeding for favourable traits may be carried out on a large scale in a very short space of time. Though they have been studied for at least a century, only recently have researchers made great leaps in discoveries.

HERPETOLOGY

Herpetology is the scientific study of amphibians and reptiles. Like most other fields of vertebrate biology (e.g., ichthyology, mammalogy), herpetology is composed of a number of cross-disciplines: Behavior, ecology, physiology, anatomy, paleontology, taxonomy, and others. Most students of recent forms are narrow in their interests, working on only one order or suborder (e.g., frogs, salamanders, snakes, lizards). A paleontologist is more likely to work with both amphibians and reptiles or with intermediate forms.

Herpetology as a unified science apparently stems from the ancient tendency to lump together all creeping animals. Modern herpetology is a truly popular science, in which amateurs have made many valuable contributions in such areas as distribution, Behavior, and even taxonomy. The major part of the more technical research is carried out at universities and museums, as well as in the field.

Research into the biology of different amphibians and reptiles has contributed much to the field of general biology as for example, larval frogs and salamanders in the understanding of embryological concepts, iguanid lizards with the development of the subdiscipline of population ecology, and snake venom in increasing the understanding of human cardiac and neurological disorders.

ORNITHOLOGY

Ornithology is a branch of zoology dealing with the study of birds. Most of the early writings on birds are more anecdotal than scientific, but they represent a broad foundation of knowledge, including much folklore, on which later work was based. In the European Middle Ages many treatises dealt with the practical aspects of ornithology, particularly falconry and game-bird management. From the mid-18th to the late 19th century, the major thrust was the description and classification of new species, as scientific expeditions made collections in tropical areas rich in bird species. By the early 20th century the large majority of birds were known to science, although the biology of many species was virtually unknown. In the latter half of the 19th century much study was done on the internal anatomy of birds, primarily for its application to taxonomy. Anatomical study was overshadowed in the first half of the 20th century by the rising fields of ecology and ethology (the study of Behavior) but underwent a resurgence beginning in the 1960s with more emphasis on the functional adaptations of birds.

Ornithology is one of the few scientific fields in which nonprofessionals make substantial contributions. Much research is carried out at universities and museums, which house and maintain the collections of bird skins, skeletons, and preserved specimens upon which most taxonomists and anatomists depend. Field research, on the other hand, is conducted by both professionals and amateurs, the latter providing valuable information on Behavior, ecology, distribution, and migration.

Although much information about birds is gained through simple, direct field observation (usually aided only by binoculars), some areas of ornithology have benefited greatly from the introduction of such instruments and techniques as bird banding, radar, radio transmitters (telemeters), and high-quality, portable audio equipment.

Bird banding (or ringing), first performed early in the 19th century, is now a major means of gaining information on longevity and movements. Banding systems are conducted by a number of countries, and each year hundreds of thousands of birds are marked with numbered leg bands. The study of bird movements has also been greatly aided by the use of sensitive radar. Individual bird movements are also recorded on a day-to-day basis by the use of minute radio transmitters (telemeters) worn by or implanted inside the bird. Visual markings, such as plumage dyes and plastic tags on the legs or wings, allow visual recognition of an individual bird without the difficult task of trapping it and allow the researcher to be aided by amateur bird-watchers in recovering his marked birds. Research into the nature and significance of bird calls has burgeoned with the development of high-quality, portable audio equipment.

References

- Mammalogist, careers: aboutbioscience.org, Retrieved 2 January, 2019
- Primatology, science: britannica.com, Retrieved 3 February, 2019
- Anthrozoology, science: britannica.com, Retrieved 4 March, 2019
- Ichthyology, entry: newworldencyclopedia.org, Retrieved 5 April, 2019
- Entomology: environmentalscience.org, Retrieved 6 May, 2019
- Herpetology, science: britannica.com, Retrieved 7 June, 2019
- Ornithology, science: britannica.com, Retrieved 8 July, 2019

3

Vertebrate Zoology

Vertebrate zoology is a sub-discipline of biology which studies animals that have vertebrae such as amphibians, fish, mammals, birds and reptiles. The chapter closely examines the key concepts of vertebrate zoology as well as these vertebrates to provide an extensive understanding of the subject.

VERTEBRATE

Vertebrates are also called Craniata. These are the animals of the subphylum Vertebrata, the predominant subphylum of the phylum Chordata. They have backbones, from which they derive their name. The vertebrates are also characterized by a muscular system consisting primarily of bilaterally paired masses and a central nervous system partly enclosed within the backbone.

The subphylum is one of the best known of all groups of animals. Its members include the classes Agnatha, Chondrichthyes, and Osteichthyes (all fishes); Amphibia (amphibians); Reptilia (reptiles); Aves (birds); and Mammalia (mammals).

Features

Although the vertebral column is perhaps the most obvious vertebrate feature, it was not present in the first vertebrates, which probably had only a notochord. The vertebrate has a distinct head, with a differentiated tubular brain and three pairs of sense organs (nasal, optic, and otic). The body is divided into trunk and tail regions. The presence of pharyngeal slits with gills indicates a relatively high metabolic rate. A well-developed notochord enclosed in perichordal connective tissue, with a tubular spinal cord in a connective tissue canal above it, is flanked by a number of segmented muscle masses. A sensory ganglion develops on the dorsal root of the spinal nerve, and segmental autonomic ganglia grow below the notochord. The trunk region is filled with a large, bilateral body cavity (coelom) with contained viscera, and this coelom extends anteriorly into the visceral arches. A digestive system consists of an esophagus extending from the pharynx to the stomach and a gut from the stomach to the anus. A distinct heart, anteroventral to the liver, is enclosed in a pericardial sac. A basic pattern of closed circulatory vessels is largely preserved in most living forms. Unique, bilateral kidneys lie retroperitoneally (dorsal to the main body cavity) and serve blood maintenance and excretory functions. Reproductive organs are formed from tissue adjacent to the kidneys; this original close association is attested by the tubular connections seen in males of living forms. The ducts of the excretory organs open through the body wall into a cloacal chamber, as does the anus of the digestive tract. Reproductive cells are shed through nearby abdominal pores or through special ducts. A muscular tail continues the axial musculature of the trunk.

Approximately 45,000 living species constitute the vertebrates. Species of several classes are found from the high Arctic or Antarctic to the tropics around the Earth; they are missing only from interior Antarctica and Greenland and from the North Polar ice pack. In size, vertebrates range from minute fishes to elephants and whales (of up to 100 tons), the largest animals ever to have existed. Vertebrates are adapted to life underground, on the surface, and in the air. They feed upon plants, invertebrate animals, and one another. Vertebrate faunas are important to humans for food and recreation.

The vertebrates are subdivided here into major groups based on morphology: the cyclostomes (jawless fishes), the chondrichthyes (cartilaginous fishes), the teleostomes (bony fishes), and the tetrapods.

The Cyclostomes

The cyclostomes include two classes of living, jawless fishes (agnathous)—Petromyzontiformes (lamprey eels) and Myxiniformes (hagfishes). The hagfishes are totally marine, often living in deep waters associated with muddy bottoms. The lampreys may be marine as adults but spawn in fresh waters, where the larvae spend some time before metamorphosing to the adult. Some lampreys live entirely in fresh water and may change only slightly in habit as a result of metamorphosis. Without lateral fins, lampreys swim by undulations of the body and can control direction only for short distances.

Lamprey (Lampetra) on rainbow trout.

The living agnaths are predatory, the lampreys being well known for attacking salmonoid fishes. The lamprey attaches to its prey using its round, suctorial mouth, and it rasps a hole through the outer tissues using a tongue armed with keratinized teeth. It suctions off bits of tissue, blood, and body fluids. The hagfishes feed somewhat similarly, but on a variety of prey—invertebrates (worms and soft-bodied forms) and dead fishes.

The lampreys produce small eggs, which develop directly into larvae that burrow into the muddy bottom of the stream. With its mouth at the surface of the mud, the larva filter feeds until large enough to metamorphose and swim off as a small adult. In contrast, the hagfishes produce relatively large encapsulated, yolky eggs up to two centimetres in length. When laid, these eggs attach to any available object by terminal hooks. The encased egg develops more or less directly into a miniature adult.

The Chondrichthyes

The sharks, rays, and chimaerids are usually marine, but some sharks have entered fresh waters (the Amazon) or even live there permanently (Lake Nicaragua). In size, sharks range from the whale shark, nearly 10 metres in length, to rather small species, three centimetres in length. They usually weigh 25 to 200 kilograms (55 to 440 pounds). Sharks are predatory animals. Some large shark species (basking and whale sharks) filter feed on small crustaceans. Herbivorous sharks are unknown. Sharks swim by undulations of the tail, but rays "fly" through the water by undulations of the pectoral fins. Most species occur in near-shore waters, but some range widely throughout the oceans. A few are found in deep water.

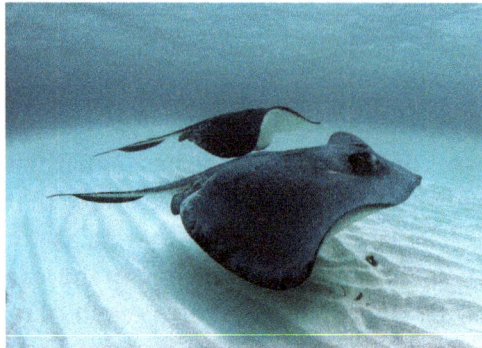
Southern stingrays (Dasyatis americana).

A few sharks produce live young (viviparous) after internal fertilization. The posterior angle of the male's pelvic fins are modified into a clasper, which acts as an intromittent organ in copulating with the female. Most sharks lay large yolky, encapsulated eggs with hooks for attachment. The young develop directly and begin life as miniature adults. The young that develop in the mother's uterus obtain nutrients from the large yolk sac until they are born alive. In a few cases, the uterine wall secretes nutrients.

Bull shark (Carcharhinus leucas).

The teleostome, or osteichthyian, fishes (those having an internal bony skeleton) can be divided into two groups: the subclasses Actinopterygii (ray-finned fishes) and Sarcopterygii (lobe-finned fishes). The latter group includes the lungfishes, which live in marshes, ponds, or streams, and are frequent air breathers. They lay fairly large eggs, with a limited amount of yolk, that are enclosed in jelly coats like those of an amphibian. The eggs develop into small fishes that feed on live prey. The larvae of the African lungfish have external gills to supplement oxygen intake.

The Teleostomes

Actinopterygian fishes are the common bony fishes of modern aquatic environments. They range in size from fishes that are only millimetres in size to those two or more metres (6.6 or more feet) in length, weighing 500 kilograms or more. Large species (sturgeons) are found in fresh waters (several other large species are found in the Amazon) as well as in marine environments. The diet may include plants, animals, and carrion. Most species are midwater swimmers, but many spend much time lying on the bottom. Tail, pectoral, and even dorsal fins are used in swimming. Reproduction in this group is by way of large numbers of small eggs, which produce small larvae or develop directly to the adult.

American paddlefish (Polyodon spathula).

Common mola, or ocean sunfish (Mola mola).

The Tetrapods

The tetrapods live primarily on land and are rather similar in habit. Members include the amphibians, reptiles, birds, and mammals. Amphibians are widespread in the warmer parts of the continents, being absent only in the far north and in the Antarctic. Three orders are recognized: Candata (the salamanders), the frogs and toads (Anura, or Salientia), and the Apoda or Gymnophiona (caecilians). Modification takes many forms, from the moist glandular skin (some scale remnants persist in apodans) to the loss of many of the bones of the skull. Like their ancestors, amphibians are cold-blooded and tend to be aquatic or limited to moist surroundings. Salamanders are seemingly the least modified in body form. They do not actively pursue prey and at best are only marginal swimmers. In swimming or crawling, the salamander's body and tail undulate. Frogs and toads hop using hindlimb propulsion and the forelimbs as body props. This dominance of the hind limb in locomotion is best seen in swimming when the forelimbs are drawn back against the body. In contrast to the salamanders and frogs, the burrowing, wormlike apodans are without limbs.

European pond turtle (Emys orbicularis).

Sandhill cranes (Grus canadensis).

Amphibians usually trap food using a tongue that can be shot out of the mouth, or they use the mouth itself to grasp and ingest food. There is great variation in foods; only the larvae of frogs and toads appear to be plant feeders, a specialization that is reflected in the highly modified jaws and guts of the tadpoles.

Amphibians have retained a simple egg cell with a gelatinous cover. The eggs are laid in ponds, streams, or even in damp places high in trees, usually in great numbers. Fertilized eggs develop into free-swimming larvae, which then metamorphose to adults, but in highly specialized forms.

The class Reptilia retains many of the structural characteristics of the ancestral amphibian. While most reptiles are carnivorous, feeding on other organisms, a few are herbivorous (e.g., tortoises). As cold-blooded animals, reptiles tend to be limited to temperate and tropical areas, but, where found, they are relatively common, although not as large or conspicuous as birds or mammals. Most reptiles are terrestrial, but a few are aquatic. As basic tetrapods, reptiles move about by creeping or swimming in a fashion similar to amphibians. Some reptiles, however, can lift the body from the ground and run rapidly either in a quadrupedal or bipedal fashion. Reptiles lay relatively large, shelled eggs. In a few instances, the eggs and young are cared for by the female; in others, the young are born alive (ovovivipary).

Birds are warm-blooded, and, although most are capable of flight, others are sedentary and some are flightless. Like their relatives the reptiles, birds lay shelled eggs that differ largely in the amount of calcification (hardening) of the shell. The young are usually cared for in a nest until they are capable of flight and self-feeding, but some birds hatch in a well-developed state that allows them to begin feeding immediately or even take flight. The megapods lay their eggs in mounds of rotting vegetation, which supplies the heat for incubation. (Nesting activities similar to those of some birds are seen in the crocodilians.)

The mammals range in size from tiny shrews or small bats weighing only a few grams to the largest known animals, the whales. Most mammals are terrestrial, feeding on both animal and vegetable matter, but a few are partially aquatic or entirely so, as in the case of the whales or porpoises. Mammals move about in a great variety of ways: burrowing, bipedal or tetrapedal running, flying, or swimming. Reproduction in mammals is usually viviparous, the young developing in the uterus, where nutritive materials are made available through an allantoic placenta or, in a few cases, a yolk sac. The fertilized egg develops directly into the adult. The monotremes (platypus and echidna) differ from other mammals in that they lay eggs which hatch, and the relatively undeveloped young are carried in a pouch or kept in a nest; the growing young lap up a milk nutrient fluid exuded from the belly of the mother.

Form and Function

External Features

The evolution of the notochord, dorsal nerve tube, and pharyngeal slits in chordate structure suggests improved swimming capability and probably greater ability to capture prey. Specialization in the vertebrate for the active capture of larger prey is evident both in the structure of the mouth and in the relatively simple structure of the pharynx, with its strong gill development. Specialization for feeding is again seen in the two basic groups of vertebrates, the agnathans and gnathostomes. Swimming adaptations are also numerous and involve variations both in body form and in medial fins and the two pairs of lateral fins.

Internal Features

The Skeletal System

Support and protection are provided by the exoskeletal and endoskeletal divisions of the skeletal system. The exoskeleton, when present, is basically protective but functions in tooth support in the mouth region. The endoskeleton protects the brain and spinal cord and assists primarily with locomotion in the trunk and tail regions. The endoskeleton begins as cartilage and may remain so or may develop into bone. The cartilaginous endoskeleton, found in the shark or chimaerid, is usually calcified so as to be stiffer and stronger. Bone is distinctive but highly variable; some types of bone contain cells, others do not, or the bone may be laminar, spongy, or arranged in sheathing layers around blood channels.

Tissues and Muscles

Tissue development in the vertebrate is unique in its complexity; tissues in the strict sense (defined as a mass or sheet of similar cells with a similar function), however, do not exist. The simplest situation is seen in the epidermis, but even here there is a layered system in which different cell types provide different functions (such as protection and secretion). The stratified epithelium of the vertebrate is highly characteristic of that group (a similar one is seen in only one invertebrate group, the class Chaetognatha).

Other tissues of the vertebrate are more complex than the epithelium. For example, skeletal muscle consists not only of striated muscle fibres but also of connective tissue, which binds it together and attaches it by way of tendons. This contractible tissue includes nerves and blood vessels and their contained blood. Skeletal muscles thus appear as simple organs, just as do the smooth muscles in the wall of the gut or the iris muscles of the eye. Such unique histological complexity runs through the entire body of the vertebrate.

Nervous System and Organs of Sensation

The dorsal position, tubular structure, and epidermal origin of the central nervous system are definitive of the chordates, although some may see similarities with the hemichordates. The sensory structures are distinctive of the chordates and include the paired nasal, optic, and otic organs (along with the strongly differentiated head).

The nasal vesicle is variously open to the environment, and its sensory cells, as chemical receptors, are not unlike those in the taste buds of the mouth. The eye is the most complex organ of the head and is a lateral outpocketing of the anterior end of the brain tube. Later it acquires a lens of epidermal origin. The act of focusing the eye (accommodation) shows extensive adaptive variation among the different groups of vertebrates.

The otic vesicle starts from a simple sac formed by the invagination of an ectodermal placode. These developmental changes also include the changes of innervation. Whereas the original structure was basically an equilibrium adaptation, other functions, such as an awareness of movement or the sensation of the proximity of prey, developed.

The lateral-line system of canals and sensory organs is a unique vertebrate feature. The elements of this system are found on the head as well as the body. This system is related to the ear and presumably at its origin served a similar function. This system is lost in terrestrial vertebrate forms.

The Digestive System

The digestive system of the vertebrate is distinctive in its structure but not in its function. The mouth and pharynx can be considered as parts of this system; the latter as an expanded cavity in the head is unmatched in any other group.

Presumably the original condition of the digestive glands was that of a ventral diverticulum which may have received the food mass into its cavity. This diverticulum, matched by the diverticulum seen in the amphioxus or the "intestine" of the tunicate, produced the secretions (bilelike and enzymes) of both liver and pancreas. Through time, the liver gradually differentiated from the pancreas. The size and separation of the liver from the gut suggest its separate blood and metabolic activities. The most obvious by-product of the liver, bile, necessitated the formation of a gall bladder and a duct connection with the gut. The pancreas, in contrast, continued to produce digestive enzymes, but its secretory cells were no longer in direct contact with the food mass. Because the pancreas was only a partial source of intestinal digestive enzymes, it was sometimes reduced in size and enclosed in the gut wall itself (agnaths) or dispersed as tiny bits of tissue in the mesentary supporting the gut (actinopterygians).

The Excretory System

The excretory system is unique in its nephrons, which filter the blood in the glomeruli and remove a variety of wastes from the body through selective secretion and reabsorption. In the shark or the coelacanth Latimeria, urea is used to raise the osmotic pressure of the blood to that of the marine habitat, thus saving these organisms considerable metabolic energy. The large intestine (sometimes centred in a rectal gland) acts as an auxiliary excretory organ, as do also the gills of fishes or the sweat glands of mammals.

Respiration and Gas Exchange

Respiration, like excretion, involves specialized body structures, such as lungs or gills, but also can involve other areas, such as the skin itself. Respiration involves exchange of gases both between the body of the organism and the environment and between the blood system and the body tissues. It also involves cellular respiration where oxygen is used and carbon dioxide is produced. There is nothing characteristic of the vertebrate in this functional area; even the hemoglobin of the blood is suggested in the respiratory pigments of other animals.

The Circulatory System

The circulatory system of vertebrates is closed in that fluids course through vessels, but there is free movement of cells in and out of blood. Some leukocyte (white blood cell) movement out of the capillaries and fluid leakage are observed in all tissues. Blood tissues are distinctive in the range of specialized cells, although these vary in detail among animals. The immune function of the blood is best developed in the vertebrate.

The Endocrine System

The endocrine system is characterized by its separate organs. The occurrence of a pituitary or a thyroid gland is suggestive of the evolutionary change and specialization that took place within this group. The relatively unspecialized nature of some parts of this system is seen in certain scattered cells in the gut wall or even the clumps of islet cells of the pancreas.

Evolution and Paleontology

The knowledge of vertebrates as revealed by fossils has grown rapidly during the past few decades, but there is much still to be discovered. The ancestral vertebrate (protovertebrate) has been sought for more than 100 years, and the likelihood of finding it today is not much greater than in the past. It can be assumed that the protovertebrate was small and soft-bodied, two factors that suggest the improbability of finding a fossilized form in a recognizable condition. There are Cambrian fossils that have been suggested to be fossil cephalochordates and there are scales of agnath fishes, but the first type of fossil is too simple and the second already too complex to explain the transition.

Annotated Classification

Subphylum Vertebrata (or Craniata)

Bilaterally symmetrical; internal skeletal support with skull enclosing a highly developed brain and a vertebral column and nerve cord; paired, jointed appendages; skin; advanced organ systems; sense organs concentrated in head.

- Class Agnatha (hagfishes, lampreys)

  Primitive; jawless; paired fins are poorly developed or lacking; rasping tongue; notochord without bone; skin is soft, glandular, and slimy; true gill arches absent; marine habitat.

- Class Placodermi (placoderms)

  †Extinct; fishlike; jaws supported by both cranium and hyoid arch (amphistylic); partly ossified cranium; primitive; head and trunk have armour that is jointed at the neck; pelvic fins present or absent; pectoral fins or finlike structures often present; gill arches.

- Class Chondrichthyes (sharks, rays, and skates)

  Cartilaginous fishes; jaws; paired fins; no swim bladder; pelvic fins in males often modified to form claspers; gill arches internal to gills; reduced notochord; lateral-line system; paired nostrils; internal nares absent; separate sexes; internal fertilization and direct development; oviparous, ovoviviparous, or viviparous.

- Subclass Elasmobranchii (sharks and rays)

  Numerous teeth derived of placoid scales; 5 to 7 gill clefts; operculum absent; cloaca; upper jaw not fused with braincase; dorsal fin nonerectile; with spiracles; worldwide distribution.

- Subclass Holocephali (chimeras)

  Teeth fused to bony plates; no scales; 4 gill pairs under 1 gill opening on each side; no cloaca; no spiracles; operculum present; upper jaw fused to braincase; dorsal fin erectile; whiplike tail; claspers present in males; temperate marine freshwater.

- Class Osteichthyes (bony fishes)

  Jaws; partly or fully ossified skeleton; usually a swim bladder; paired fins; gills covered by a bony operculum; scales; paired nostrils with or without internal nares; lateral-line system; mostly oviparous with external fertilization; some ovoviviparous or viviparous.

- Subclass Actinopterygii (ray-finned fishes)

  Generally lack choanae; no fleshy base to paired fins; no internal nares; air sacs usually function as swim bladder; skeleton usually well ossified.

- Subclass Sarcopterygii (lobe-finned fishes)

  Usually possess a choana; paired fins with a fleshy base over a bony skeleton; persisting notochord; 2 dorsal fins; nares are internal.

- Class Amphibia

  Cold-blooded; respire by lungs, gills, skin, or mouth lining; larval stage in water or in egg; skin is usually moist with mucous glands and without scales; tetrapods; freshwater and terrestrial; paired appendages are legs; 10 pairs of cranial nerves; separate sexes; external fertilization with development into tadpole larvae; some have internal development, ovoviviparous or viviparous.

- Order Aponda (or Gymnophiona; caecilians)

  Wormlike; no limbs or girdles; compact skull; lidless, minute eyes; persistent notochord; tail; scales present in some species.

- Order Anura (or Salientia; frogs and toads)

  Tailless; elongated hind limbs modified for jumping; larvae lack true teeth and external gills.

- Order Caudata (or Urodela; salamanders)

  Tail; limbs normal; many skeletal elements cartilaginous; larvae with true teeth and external gills.

- Class Reptilia

  Cold-blooded; no larval stage; breathing by lungs; well-ossified skull; dry skin; scales; no glands; 5-toed limbs; claws; 3- or 4-chambered heart with incomplete ventricle separation; 12 pairs of cranial nerves; internal fertilization, direct development; oviparous and ovoviviparous.

- Subclass Anapsida (turtles, tortoises, terrapins)

  No temporal skull openings; body encased in bony shell; no teeth in living members; oviparous.

- Subclass Lepidosauria

  No bipedal specializations; 2 complete temporal openings; complete palate; oviparous; male is without penis.

- Subclass Archosauria (ruling reptiles)

  Some ancient forms had bipedal locomotion; longer hind legs; semiaquatic; webbed feet; teeth in sockets; single penis; oviparous; includes extinct dinosaurs.

- Subclass Synaptosauria

  Extinct; single temporal opening on area of cheek.

- Subclass Ichthyopterygia

  Extinct; temporal openings high up on skull; fishlike; spindle-shaped body; high tail fin; triangular dorsal fin; paddlelike legs; marine.

- Subclass Synapsida

  Extinct; mammallike; lateral temporal opening.

- Class Aves

  Warm-blooded; skull has only 1 condyle; front limbs primarily modified for flight; hind limbs are legs with 4 or fewer toes; body covered with feathers; scales on feet; 4-chambered heart; no teeth; horny beak; lungs with extended air sacs; 12 pairs of cranial nerves; internal fertilization; oviparous.

- Subclass Archaeornithes

  Extinct; teeth in both jaws; long, feathered tail; less specialized for flight; body elongated and reptilelike; forelimb had 3 clawed digits; small brain and eyes; nonpneumatic bones.

- Subclass Neornithes (true birds)

  Well-developed sternum; tail is not long; no teeth; forelimbs modified to wings; teeth replaced by horny rhamphoteca over bill.

- Class Mammalia

  Warm-blooded; mammary glands; lower jaw is composed of 1 bone; hair; advanced brain; skin with different glands and hair; ears with 3 middle-ear bones; 12 pairs of cranial nerves; 4-chambered heart; young nourished by milk from mammary gland; internal fertilization; mostly viviparous, some oviparous.

- Subclass Prototheria

  Primitive; egg-laying; hair; mammary glands without nipples; pectoral girdle; separate oviducts that open into cloacal chamber that is shared with excretory ducts; oviparous.

- Subclass Theria

  Mammary glands with nipples; functional teeth; oviducts partly fused; with or without a cloaca; uterus and vagina; viviparous.

Critical Appraisal

The classification of animals is presently in a state of flux. The classification presented here is traditional and conservative. Because traditional theories of taxonomy tend to be nonquantitative, various interpretations of relationships or patterns can be presented and defended.

The alternative cladistic style of taxonomy is an attempt to force taxonomy into a testable, highly objective operation. One tentative classification based in cladistics separates the vertebrates into two superclasses (Agnatha and Gnathostomata). Agnathans are jawless, while the gnathostomates encompass the remainder of the jawed vertebrates. Living agnathans are placed in the class Cyclostomata. Gnathostomates can be further divided into the epiclasses Elasmobranchiomorphi (sharks and rays) and Teleostomi (bony fishes and tetrapods). The former group are identified primarily by a cartilaginous skeleton, while the latter group have developed a bony skeleton. Two subepiclasses of the teleostomes are Ichthyopterygii (or Osteichthyes; bony fishes) and Cheiropterygii (tetrapods), the latter being further divided into the classes Amphibia, Reptilia, Aves, and Mammalia.

Although this classification includes and uses traditional taxonomic categories, their position in the hierarchy may be changed. Separation of agnath and gnathostome is opposed by those cladists who chart the origin of gnathostomes from the agnath, believing that the differences in mouth and tooth structure are a result of modification. The Gnathostomata is subdivided into the Elasmobranchiomorphi and the Teleostomi largely on the basis of mouth and tooth structure. The creation of epiclasses and subepiclasses in the alternative classification is not important in itself; the creation of a dichotomy between Ichthyopterygii and Cheiropterygii, however, is important, although from the evolutionary view it is evident that the one evolved from the other.

AMPHIBIANS

Amphibians are cold-blooded vertebrates that include the well-known frogs and toads. Being cold-blooded means that they depend on environmental sources of heat to regulate their body heat and temperature. The class of Amphibia is made up of more than 3,500 species which include the various order of amphibians. Most amphibians begin their lives in water and eventually adapt to life on land by developing lungs and limbs that allow them to move on land. The larvae mature while in the water. At this young stage, the offspring breathe through the gills and after some time they develop lungs through a process known as metamorphosis. The class of Amphibia is made of three orders namely; Anura (toads and frogs), Urodela or Caudata (newts and salamanders), and Apoda or Gymnophiona (caecilians).

Major Characteristics of Amphibians

Amphibians have characteristics that cross-over between fish and reptiles. At the youngster age, most of them function like fish while as adults, they have different characteristics that allow them to live on land. They are cold-blooded animals which regulate their body heat and temperature depending on the external environment.

Amphibians have scale-less skin that is very delicate and moist. They live close to water sources in order to dampen their skin. The skin greatly helps in regulating the body temperature but also makes them vulnerable to dehydration. In high temperatures, dehydration will lead to death. This is the reason why amphibians live close to marshes, swamps and ponds and other freshwater bodies.

They breathe oxygen through the skin. The skin plays an important role in gas exchange and in absorption of water. This is despite them having lungs which function rather poorly under certain conditions. The skin, therefore, plays a double role of protecting and absorbing the water and oxygen.

Certain frogs such as the brightly colored poison-dart frog, have skin that contains poisonous glands. The poison is used as a defense mechanism which can easily kill any predator or prey. The poison from frogs has been used by Native American Indians hunters to coat the tips of their spears and arrows.

Three Orders of Amphibians

All amphibians are classified according to bodily characteristics of their legs and tails.

Anura

Anura is the largest order of living amphibians with over 3,000 different varieties. Toads and frogs fall under the order of Anura. This group lacks a tail and are characterized by long hind limbs that are adapted for swimming and leaping. Anura amphibians live in freshwater regions although some may be found in drier habitats. Frogs and toads are different in their body characteristic. Toads usually have shorter hind limbs and drier skin that appears warty, while frogs have a thin smooth skin and long hinder limbs. Anura amphibians feed on a variety of invertebrates such as insects. They can also feed on small mammals, birds, and fish.

Urodeles

Newts and salamanders fall under this category. The largest amphibian, the Japanese salamander, measures up to 1.5 meters while the smallest member of this order measures 10 centimeters in length. In this order, the tail is more pronounced than the limbs which are usually underdeveloped. Their preferred habitat is near water bodies and under moist soil and rocks. They mostly feed on insects and worms. Some species live in water, such as the genus Siren, while others burrow in the mud. They have lungs and external gills to aid in breathing.

Apoda

Apoda consists of about 205 species. They are shaped like worms, legless, and blind. They can be found in mud where they live, especially in the tropical soils of Africa and South America. They measure between 10 centimeters and 1 meter in length.

Amphibian Life Cycle

The life of amphibians begins in water where the female lays eggs that are externally fertilized. After the eggs hatch into tadpoles, they breathe through external gills. Tadpoles have flat tails that are used for swimming, and feed on aquatic vegetation. Eventually, through metamorphosis, they experience physical changes that make them adults. This includes the developing of lungs and elaborate limbs that aid them in movement on land.

Important Roles of Amphibians

Amphibians such as frogs are vital to the balance of ecosystem in which they inhabit, both as predators or prey. They feed on pests and insects thereby reducing the spread of diseases to agricultural plants. This indirectly benefits agriculture. In certain cultures around the world, frogs are viewed as a source of luck and are cherished as important symbols in society. In medical research, the skin of amphibians is being studied due to their ability to resist virus infections. This could eventually provide an advance in the treatment of virus diseases such as AIDS.

Main Threats to the Existence of Amphibians

Today, the number of amphibian species has continued to decline due to a variety of reasons. This includes the pollution of freshwater ecosystems which provide habitat for most of the species. Ultraviolet radiation has also affected the thriving of amphibians due to their fragile skin. Additionally, diseases such as the Chytrid fungus have depopulated many amphibian habitats. Many have been wiped out at a rapid rate to the extent that they are not even noticeable. The loss of amphibians affects the balance of the ecosystem which in turn affects other animal and plant species on the planet.

MAMMAL

Mammal (class Mammalia), is the member of the group of vertebrate animals in which the young are nourished with milk from special mammary glands of the mother. In addition to these characteristic milk glands, mammals are distinguished by several other unique features. Hair is a typical mammalian feature, although in many whales it has disappeared except in the fetal stage. The mammalian lower jaw is hinged directly to the skull, instead of through a separate bone (the quadrate) as in all other vertebrates. A chain of three tiny bones transmits sound waves across the middle ear. A muscular diaphragm separates the heart and the lungs from the abdominal cavity. Only the left aortic arch persists. (In birds the right aortic arch persists; in reptiles, amphibians, and fishes both arches are retained.) Mature red blood cells (erythrocytes) in all mammals lack a nucleus; all other vertebrates have nucleated red blood cells.

Except for the monotremes (an egg-laying order of mammals comprising echidnas and the duck-billed platypus), all mammals are viviparous—they bear live young. In the placental mammals (which have a placenta to facilitate nutrient and waste exchange between the mother and the developing fetus), the young are carried within the mother's womb, reaching a relatively advanced stage of development before birth. In the marsupials (e.g., kangaroos, opossums, and wallabies), the newborns are incompletely developed at birth and continue to develop outside the womb, attaching themselves to the female's body in the area of her mammary glands. Some marsupials have a pouchlike structure or fold, the marsupium, that shelters the suckling young.

The class Mammalia is worldwide in distribution. It has been said that mammals have a wider distribution and are more adaptable than any other single class of animals, with the exception of certain less-complex forms such as arachnids and insects. This versatility in exploiting Earth is attributed in large part to the ability of mammals to regulate their body temperatures and internal environment both in excessive heat and aridity and in severe cold.

Features

Diversity

The evolution of the class Mammalia has produced tremendous diversity in form and habit. Living kinds range in size from a bat weighing less than a gram and tiny shrews weighing but a few grams to the largest animal that has ever lived, the blue whale, which reaches a length of more than 30 metres (100 feet) and a weight of 180 metric tons (nearly 200 short [U.S.] tons). Every major habitat has been exploited by mammals that swim, fly, run, burrow, glide, or climb.

Okapi (Okapia johnstoni).

There are more than 5,500 species of living mammals, arranged in about 125 families and as many as 27–29 orders (familial and ordinal groupings sometimes vary among authorities). The rodents (order Rodentia) are the most numerous of existing mammals, in both number of species and number of individuals, and are one of the most diverse of living lineages. In contrast, the order Tubulidentata is represented by a single living species, the aardvark. The Uranotheria (elephants and their kin) and Perissodactyla (horses, rhinoceroses, and their kin) are examples of orders in which far greater diversity occurred in the late Paleogene and Neogene periods (about 30 million to about 3 million years ago) than today.

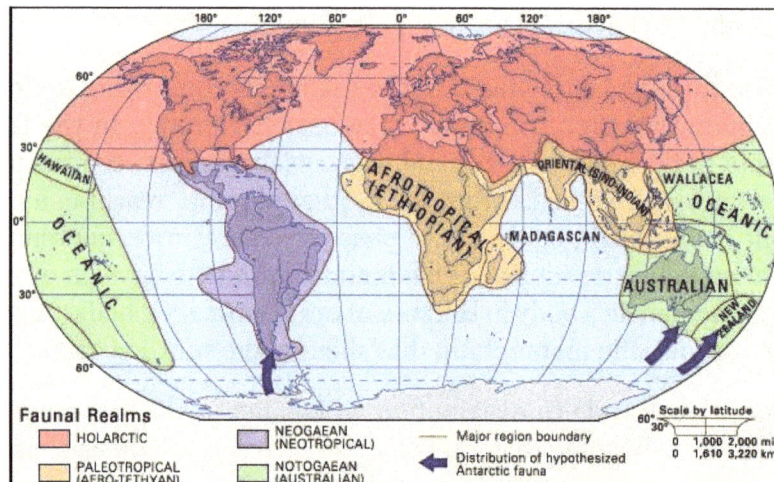

Faunal realms and major regions of the world.

The greatest present-day diversity is seen in continental tropical regions, although members of the class Mammalia live on (or in seas adjacent to) all major landmasses. Mammals can also be found on many oceanic islands, which are principally, but by no means exclusively, inhabited by bats. Major regional faunas can be identified; these resulted in large part from evolution in comparative isolation of stocks of early mammals that reached these areas. South America (the Neotropics), for example, was separated from North America (the Nearctic) from about 65 million to 2.5 million years ago. Mammalian groups that had reached South America before the break between the continents, or some that "island-hopped" after the break, evolved independently from relatives that remained in North America. Some of the latter became extinct as the result of competition with more advanced groups, whereas those in South America flourished, some radiating to the extent that they have successfully competed with invaders since the rejoining of the two continents. Australia provides a parallel case of early isolation and adaptive radiation of mammals (specifically the monotremes and marsupials), although it differs in that Australia was not later connected to any other landmass. The placental mammals that reached Australia (rodents and bats) evidently did so by island-hopping long after the adaptive radiation of the mammals isolated early on.

In contrast, North America and Eurasia (the Palearctic) are separate landmasses but have closely related faunas as the result of having been connected several times during the Pleistocene Epoch (2.6 million to 11,700 years ago) and earlier across the Bering Strait. Their faunas frequently are thought of as representing not two distinct units but one, related to such a degree that a single name, Holarctic, is applied to it.

Importance to Humans

Wild and domesticated mammals are so interlocked with our political and social history that it is impractical to attempt to assess the relationship in precise economic terms. Throughout our own evolution, for example, humans have depended on other mammals for food and clothing. Domestication of mammals helped to provide a source of protein for ever-increasing human populations and provided means of transportation and heavy work as well. Today, domesticated strains of the house mouse, European rabbit, guinea pig, hamster, gerbil, and other species provide much-needed laboratory subjects for the study of human-related physiology, psychology, and a variety of diseases from dental caries to cancer. The study of nonhuman primates (monkeys and apes) has opened broad new areas of research relevant to human welfare. The care of domestic and captive mammals is, of course, the basis for the practice of veterinary medicine.

Herding goats along the ancient Silk Road, northern Takla Makan Desert.

Wild mammals are a major source of food in some parts of the world, and many different kinds, from fruit bats and armadillos to whales, are captured and eaten by various cultural groups. In addition, hunting, primarily for sport, of various rodents, lagomorphs, carnivores, and ungulates is a multibillion-dollar enterprise. In the United States alone, for example, it is estimated that more than two million deer are harvested annually by licensed hunters.

Geopolitically, the quest for marine mammals was responsible for the charting of a number of areas in both Arctic and Antarctic regions. The presence of terrestrial furbearers, particularly beavers and several species of mustelid carnivores (e.g., marten and fisher), was one of the principal motivations for the opening of the American West, Alaska, and the Siberian taiga. Ranch-raised animals such as the mink, fox, and chinchilla are also important to the fur industry, which directly and indirectly accounts for many millions of dollars in revenue each year in North America alone.

Aside from pelts and meat, special parts of some mammals regularly have been sought for their special attributes. Rhinoceros horn is used for concocting potions in eastern Asia; ivory from elephants and walruses is highly prized; and ambergris, a substance regurgitated by sperm whales, was once widely used as a base for perfumes.

Some mammals are directly detrimental to human activities. House rats and mice of Old World origin now occur virtually throughout the world and each year cause substantial damage and economic loss. Herbivorous mammals may eat or trample crops and compete with livestock for food, and native carnivores sometimes prey on domestic herds. Large sums are spent annually to control populations of "undesirable" wild mammals, a practice long deplored by conservationists. Not only do they have an impact on food resources, but mammals are also important reservoirs or agents of transmission of a variety of diseases that afflict man, such as plague, tularemia, yellow fever, rabies, leptospirosis, Lyme disease, hemorrhagic fevers such as Ebola, and Rocky Mountain spotted fever. The annual "economic debt" resulting from mammal-borne diseases that affect humans and domestic animals is incalculable.

Many large mammals have been extirpated entirely or exist today only in parks and zoos; others are in danger of extinction, and their plight is receiving increased attention from a number of conservation agencies. By the early 21st century, the International Union for Conservation of Nature (IUCN) reported that nearly one-quarter of all mammals are at risk of extinction. The single greatest threat to these mammals is the continued destruction of their habitat; however, many species are also aggressively hunted. The IUCN classifies each imperiled mammal into one of the following categories: near threatened, vulnerable, endangered, critically endangered, critically endangered and possibly extinct, or extinct in the wild.

One of the most noteworthy cases of direct extirpation by man is the Steller's sea cow (Hydrodamalis gigas). These large (up to 10 metres, or 33 feet, long), inoffensive marine mammals evidently lived only along the coasts and shallow bays of the Komandor Islands in the Bering Sea. Discovered in 1741, they were easily killed by Russian sealers and traders for food, their meat being highly prized, and the last known live individual was taken in 1768.

The Steller's sea cow (Hydrodamalis gigas)century, fed on kelp growing near the shore.

Of final note is the aesthetic value of wild mammals and the relatively recent expense of considerable energy and resources to study and, if possible, conserve vanishing species, to set aside natural areas where native floral and faunal elements can exist in an otherwise highly agriculturalized or industrialized society, and to establish modern zoological parks and gardens. Such outdoor "laboratories" attract millions of visitors annually and will provide means by which present and future generations of humans can appreciate and study, in small measure at least, other kinds of mammals.

Reproduction

Estrus and Other Cycles

In reproductively mature female mammals, an interaction of hormones from the pituitary gland and the ovaries produces a phenomenon known as the estrous cycle. Estrus, or "heat," typically coincides with ovulation, and during this time the female is receptive to the male. Estrus is preceded by proestrus, during which ovarian follicles mature under the influence of a follicle-stimulating hormone from the anterior pituitary. The follicular cells produce estrogen, a hormone that stimulates proliferation of the uterine lining, or endometrium. Following ovulation, in late estrus, the ruptured ovarian follicle forms a temporary endocrine gland known as the corpus luteum. Another hormone, progesterone, secreted by the corpus luteum, causes the endometrium to become quiescent and ready for implantation of the developing egg (blastocyst), should fertilization occur. In members of the infraclass Eutheria (placental mammals), the placenta, as well as transmitting nourishment to the embryo, has an endocrine function, producing hormones that maintain the endometrium throughout gestation.

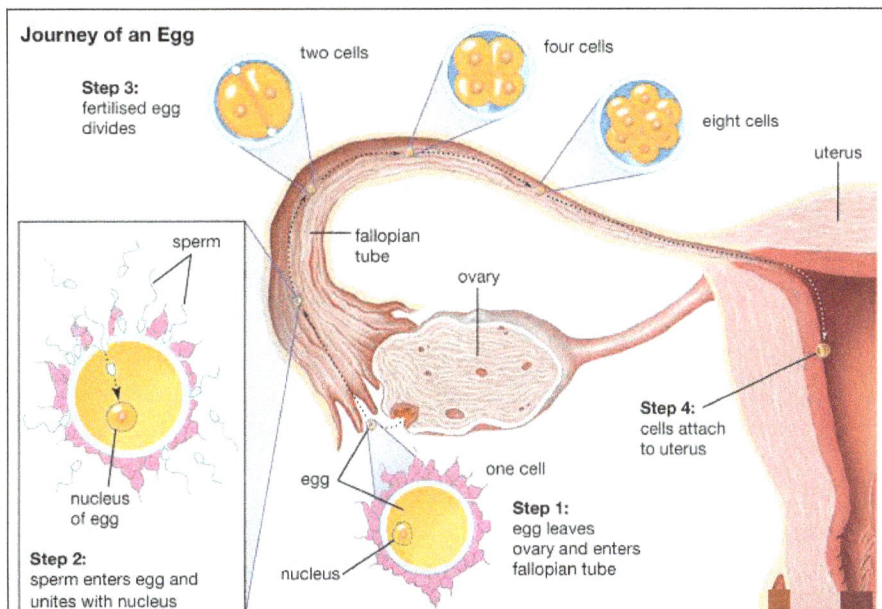

The journey of a fertilized egg in a woman. In mammals, eggs are released by the ovaries. If the egg meets a sperm cell, it may become fertilized. The fertilized egg travels to the uterus, where it grows and develops into a new individual.

If fertilization and implantation do not occur, a phase termed metestrus ensues, in which the reproductive tract assumes its normal condition. Metestrus may be followed by anestrus, a nonreproductive period characterized by quiescence or involution of the reproductive tract. On the other hand, anestrus may be followed by a brief quiescent period (diestrus) and another preparatory proestrus phase. Mammals that breed only once a year are termed monestrous and exhibit a long anestrus; those that breed more than once a year are termed polyestrous. In many polyestrous species the estrous cycle ceases during gestation and lactation (milk production), but some rodents have a postpartum estrus and mate immediately after giving birth.

The menstrual cycle of higher primates is derived from the estrous cycle but differs from estrus in that when progesterone secretion from the corpus luteum ceases, in the absence of fertilization, the uterine lining is sloughed. In anthropoids other than humans, a distinct period of "heat" occurs around the time of ovulation.

Monotremes lay shelled eggs, but the ovarian cycle is similar to that of other mammals. The eggs are predominantly yolk (telolecithal), like those of reptiles and birds. Young monotremes hatch in a relatively early stage of development and are dependent upon the parent (altricial). They reach sexual maturity in about one year.

The reproduction of marsupials differs from that of placentals in that the uterine wall is not specialized for the implantation of embryos. The period of intrauterine development varies from about 8 to 40 days. After this period the young migrate through the vagina to attach to the teats for further development. The pouch, or marsupium, is variously structured. Many species, such as kangaroos and opossums, have a single well-developed pouch; in some phalangerids (cuscuses and brush-tailed possums), the pouch is compartmented, with a single teat in each compartment. The South American caenolestids, or rat opossums, have no marsupium. The young of most marsupials depend on maternal care through the pouch for considerable periods, 13 to 14 weeks in the North American, or Virginia opossum (Didelphis virginiana). Young koalas are carried in the pouch for nearly 8 months, kangaroos to 10 months.

Implantation, Gestation and Birth

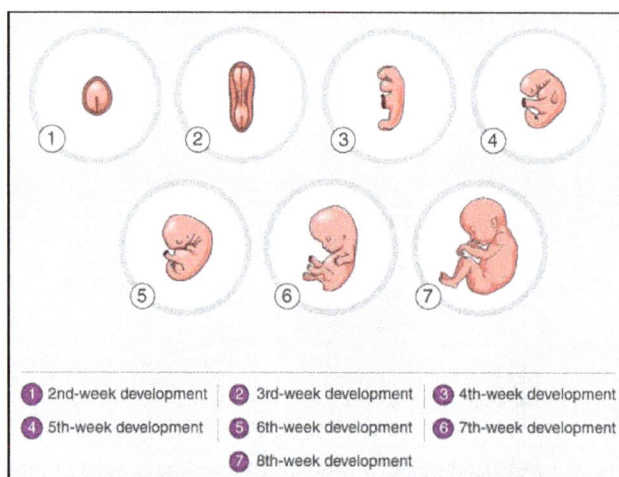

① 2nd-week development	② 3rd-week development	③ 4th-week development
④ 5th-week development	⑤ 6th-week development	⑥ 7th-week development
	⑦ 8th-week development	

The state of development of a human embryo
at the seventh week of pregnancy.

Reproductive patterns in placental mammals are diverse, but in all cases a secretory phase is present in the uterine cycle, and the endometrium is maintained by secretions of progesterone from the corpus luteum. The blastocyst implants in the uterine wall. Villi are embedded in the lining of the uterus. The resulting complex of embryonic and maternal tissues is a true placenta. The uterine lining may be shed with the fetal membranes as "afterbirth" (a condition called deciduate) or may be resorbed by the female (nondeciduate). Placentas have been classified on the basis of the relationship between maternal and embryonic tissues. In the simplest nondeciduate placental arrangement, the chorionic villi are in contact with uterine epithelium (the inner surface layer). In the "intimate deciduous" types, seen in primates, bats, insectivores, and rodents, the capillary endothelium (the layer containing minute blood vessels) of the uterine wall breaks down, and chorionic epithelium is in direct contact with maternal blood. In advanced stages of pregnancy in rabbits, even the chorionic epithelium is eroded, and the embryonic endothelium contacts the maternal blood supply. In no case, however, is there actual exchange of blood between mother and fetus; nutrients and gases must still pass through the walls of the fetal blood vessels.

The period of intrauterine development, or gestation, varies widely among eutherians, generally depending on the size of the animal but also influenced by the number of young per litter and the condition of young at birth. The gestation period of the golden hamster is about 2 weeks, whereas that of the blue whale is 11 months and that of the African elephant 21 to 22 months.

At birth the young may be well-developed and able to move about at once (precocial), or they may be blind, hairless, and essentially helpless (altricial). In general, precocial young are born after a relatively long gestation period and in a small litter. Hares and many large grazing mammals bear precocial offspring. Rabbits, carnivores, and most rodents bear altricial young.

After birth young mammals are nourished by milk secreted by the mammary glands of the female. The development of milk-producing tissue in the female mammae is triggered by conception, and the stimulation of suckling the newborn prompts copious lactation. In therians (marsupials and placentals) the glands open through specialized nipples. The newborn young of marsupials are unable to suckle, and milk is "pumped" to the young by the mother.

Milk consists of fat, protein (especially casein), and lactose (milk sugar), as well as vitamins and salts. The actual composition of milk of mammals varies widely among species. The milk of whales and seals is some 12 times as rich in fats and 4 times as rich in protein as that of domestic cows but contains almost no sugar. Milk provides an efficient energy source for the rapid growth of young mammals; the weight at birth of some marine mammals doubles in five days.

Behavior

Social Behavior

The dependence of the young mammal on its mother for nourishment has made possible a period of training. Such training permits the nongenetic transfer of information between generations. The ability of young mammals to learn from the experience of their elders has allowed a behavioral plasticity unknown in any other group of organisms and has been a primary reason for the evolutionary success of mammals. The possibility of training is one of the factors that has made increased brain complexity a selective advantage. Increased associational potential and memory extend the possibility of learning from experience, and the individual can make adaptive behavioral responses to environmental change. Individual response to short-term change is far more efficient than genetic response.

Leopard (Panthera pardus) grooming her cub in
the Masai Mara National Reserve, Kenya.

Some types of mammals are solitary except for brief periods when the female is in estrus.
Others, however, form social groups. Such groups may be reproductive or defensive, or they
may serve both functions. In those cases that have been studied in detail, a more or less strict
hierarchy of dominance prevails. Within the social group, the hierarchy may be maintained
through physical combat between individuals, but in many cases stereotyped patterns of Behavior evolve to displace actual combat, thereby conserving energy while maintaining the social structure.

A pronounced difference between sexes (sexual dimorphism) is frequently extreme in social mammals. In large part this is because dominant males tend to be those that are largest or best-armed.
Dominant males also tend to have priority in mating or may even have exclusive responsibility for
mating within a "harem." Rapid evolution of secondary sexual characteristics, including size, can
take place in a species with such a social structure.

A complex Behavior termed "play" frequently occurs between siblings, between members of an
age class, or between parent and offspring. Play extends the period of maternal training and is
especially important in social species, providing an opportunity to learn Behavior appropriate to
the maintenance of dominance.

Territoriality

Wolves are territorial mammals, and members of the pack frequently mark the limits
of the pack's territory with urine and secretions from specialized glands. Howling
also serves to alert other animals to the pack's land claims.

That area covered by an individual in its general activity is frequently termed the home range. A territory is a part of the home range defended against other members of the same species. As a generalization it may be said that territoriality is more important in the Behavior of birds than of mammals, but data for the latter are available primarily for diurnal species. Frequently territories of mammals are "marked," either with urine or with secretions of specialized glands, as in lemurs. This form of territorial labeling is less evident to humans than the singing or visual displays of birds. Many mammals that do not maintain territories per se nevertheless will not permit un-limited crowding and will fight to maintain individual distance. Such mechanisms result in more economical spacing of individuals over the available habitat.

Ecology

Response to Environmental Cycles

Mammals may react to environmental extremes with acclimatization, compensatory Behavior, or physiological specialization. One way for a mammal to endure stressful environmental conditions is to become dormant. Dormancy is the general term that relates to the reduced metabolic activity adopted by many organisms under conditions of environmental stress. Dormancy is differentiated from sleep, which is not necessarily a response to environmental stess but rather occurs as part of an organism's daily rest cycle. Physiological responses to adverse conditions include torpor, hibernation (in winter), and estivation (in summer). Torpor is a type of dormancy that may occur in the daily cycle or during unfavourable weather; short-term torpor is generally economical only for small mammals that can cool and warm rapidly. The body temperature of most temperate-zone bats drops near that of the ambient air whenever the animal sleeps. The winter dormancy of bears at high latitudes is an analogous phe-nomenon and cannot be considered true hibernation.

A cluster of Virginia big-eared bats (Corynorhinus townsendii virginianus) hibernating in Hellhole Cave, Pendleton county, West Virginia.

Strictly speaking, hibernation only occurs in warm-blooded vertebrates. True hibernation involves physiological regulation to minimize the expenditure of energy. The body temperature is lowered, and breathing may be slowed to as low as 1 percent of the rate in an active individual. There is a corresponding slowing of circulation and typically a reduction in the peripheral blood supply. When the body temperature nears the freezing point, spontaneous arousal occurs, although other kinds of stimuli generally elicit only a very slow response. In mammals that exhibit winter dorman-cy (such as bears, skunks, and raccoons), arousal may be quite rapid. Hibernation has evidently originated independently in a number of mammalian lines, and the comparative physiology of this complex phenomenon is only now beginning to be understood.

Inactivity in response to adverse summer conditions (heat, drought, lack of food) is termed estivation. Estivation in some species is simply prolonged rest, usually in a favourable microhabitat; in other species estivating mammals regulate their metabolism, although the effects are typically not as pronounced as in hibernation.

Behavioral response to adverse conditions may involve the selection or construction of a suitable microhabitat, such as the cool, moist burrows of desert rodents. Migration is a second kind of behavioral response. The most obvious kind of mammalian migration is latitudinal. Many temperate-zone bats, for example, undertake extensive migrations, although other bat species hibernate near their summer foraging grounds in caves or other equable shelters during severe weather when insects are not available. Caribou (Rangifer tarandus), or reindeer, migrate from the tundra to the forest edge in search of a suitable winter range, and a number of cetaceans (whales, dolphins, and porpoises) and pinnipeds (walruses and seals) undertake long migrations from polar waters to more temperate latitudes. Gray whales, for example, migrate southward to calving grounds along the coasts of South Korea and Baja California from summer feeding grounds in the northern Pacific Ocean (Okhotsk, Bering, and Chukchi seas). Of comparable extent is the dispersive feeding migration of the northern fur seal (Callorhinus ursinus).

Migrations of lesser extent include the elevational movements from mountains to valleys of some ungulates—the American elk (Cervus elaphus canadensis), or wapiti, and bighorn sheep (Ovis canadensis), for example—and the local migrations of certain bats from summer roosts to hibernation sites. Most migratory patterns of mammals are part of a recurrent annual cycle, but the irruptive (sudden) emigrations of lemmings and snowshoe hares are largely acyclic responses to population pressure on food supplies.

Populations

A population consists of individuals of three "ecological ages"—prereproductive, reproductive, and postreproductive. The structure and dynamics of a population depend, among other things, on the relative lengths of these ages, the rate of recruitment of individuals (either by birth or by immigration), and the rate of emigration or death. The reproductive potential of some rodents is well known; some mice are reproductively mature at four weeks of age, have gestation periods of three weeks or less, and may experience postpartum estrus, with the result that pregnancy and lactation may overlap. Litter size, moreover, may average four or more, and breeding may occur throughout the year in favourable localities. The reproductive potential of a species is, of course, a theoretical maximum that is rarely met, inasmuch as, among other reasons, a given female typically does not reproduce throughout the year. Growth of a population depends on the survival of individuals to reproductive age. The absolute age at sexual maturity ranges from less than 4 weeks in some rodents to some 15 years in the African bush elephant (Loxodonta africana).

Postreproductive individuals are rare in most mammalian populations. Survival through more than a single reproductive season is probably uncommon in many small mammals, such as mice and shrews. Larger species typically have longer life spans than do smaller kinds, but some bats are known, on the basis of banding records, to live nearly 20 years. Many species show greater longevity in captivity than in the wild. Captive echidnas are reported to have lived more than 50 years. Horses have been reported to live more than 60 years, and elephants have lived to more than 80. Various cetaceans survive to more than 90 years of age, and research involving the dating of harpoons embedded in some Greenland right whales (Balaena mysticetus), or bowheads, suggests that Greenland right whales can live 200 years or more.

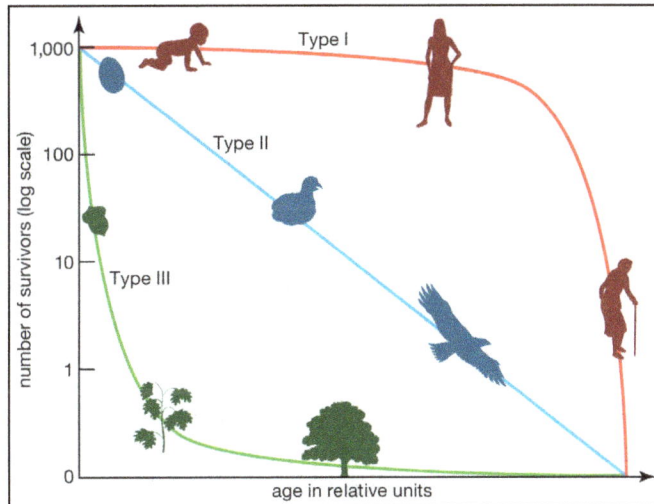

Survivorship curve Type I, II, and III survivorship curves: A survivorship curve is the graphic representation of the number of individuals in a population that can be expected to survive to any specific age.

Locomotion

Specialization in habitat preference has been accompanied by locomotor adaptations. Terrestrial mammals have a number of modes of progression. The primitive mammalian stock walked plantigrade—that is, with the digits, bones of the midfoot, and parts of the ankle and wrist in contact with the ground. The limbs of ambulatory mammals are typically mobile, capable of considerable rotation.

Brown bear (Ursus arctos) running across the water in Katmai National Park and Preserve, Alaska.

Mammals modified for running are termed cursorial. The stance of cursorial species may be digitigrade (the complete digits contacting the ground, as in dogs) or unguligrade (only tips of digits contacting the ground, as in horses). In advanced groups limb movement is forward and backward in a single plane.

Saltatory (leaping) locomotion, sometimes called "ricochetal," has arisen in several unrelated groups (some marsupials, lagomorphs, and several independent lineages of rodents). This mode of locomotion is typically found in mammals living in open habitats. Jumping mammals typically have elongate, plantigrade hind feet, reduced forelimbs, and long tails. Convergent evolution within a given adaptive mode has contributed to the ecological similarity of regional mammalian faunas.

Mammals of several orders have attained great size (elephants, hippopotamuses, and rhinoceroses) and have converged on specializations for a ponderous mode of locomotion referred to as "graviportal." These animals have no digit reduction and deploy the digits in a circle around the axis of the limb for maximum support, like the pedestal of a column.

Bats are the only truly flying mammals. Only with active flight have the resources of the aerial habitat been successfully exploited. Mammals belonging to other groups (colugos, marsupials, rodents) are adapted for gliding. A gliding habit is frequently accompanied by scansorial (climbing) locomotion. Many nongliders, such as tree squirrels, are also scansorial.

Well-adapted arboreal mammals frequently are plantigrade, five-toed, and equipped with highly mobile limbs. Some species, including many New World monkeys, have a prehensile tail, which is used like a fifth hand. Brachiation, or "arm walking," in which the animal hangs from branches and moves by a series of long swings, is an adaptation seen in gibbons. The primitive opposable anthropoid thumb is reduced as a specialization for this method of locomotion. Tarsiers are highly arboreal primates that have expanded pads on the digits to improve grasping, whereas many other arboreal mammals have claws or well-developed nails.

Several mammalian groups (sirenians, cetaceans, and pinnipeds) have independently assumed fully aquatic habits. In some cases semiaquatic mammals are relatively unmodified representatives of otherwise terrestrial groups (otters, muskrats, and water shrews, for example). Other kinds have undergone profound modification for natatorial (swimming) locomotion for life at sea. Pinniped carnivores (walruses and seals) give birth to their young on land, but cetaceans are completely helpless out of water, on which they depend for mechanical support and thermal insulation.

Food Habits

The earliest mammals, like their reptilian ancestors, were active predators. From such a basal stock there has been a complex diversification (radiation) of trophic level adaptations. Modern mammals occupy a wide spectrum of feeding niches. In most terrestrial and some aquatic communities, carnivorous mammals are the top predators. Herbaceous mammals often serve as primary consumers in most ecosystems. The voracious shrews, smallest of mammals, sometimes prey on vertebrates larger than themselves. They may eat twice their weight in food each day to maintain their active metabolism and compensate for heat loss caused by an unfavourable surface-to-volume ratio. On the other hand, the largest of vertebrates, the blue whale (Balaenoptera musculus), feeds on minute planktonic crustaceans called krill.

Within a given lineage, the adaptive radiation of food habits may be broad. Some of the carnivores have become omnivorous (raccoons, bears) or herbivorous (giant panda). Marsupials exhibit a great variety of feeding types, and in Australia marsupials have radiated to fill ecological niches highly analogous to those of placental mammals elsewhere; there are marsupial "moles," "anteaters," "mice," "rats," "cats," and "wolves." Some bandicoots have ecological roles similar to those of rabbits, and wombats are partially burrowing (semifossorial) herbivores analogous to marmots. In Australia the niche of large grazers and browsers is filled by a variety of kangaroos and wallabies.

Niche	Placental Mammals	Australian Marsupials
Burrower	Mole	Marsupial mole
Anteater	Lesser anteater	Numbat (anteater)
Mouse	Mouse	Marsupial mouse
Climber	Lemur	Spotted cuscus
Glider	Flying squirrel	Flying phalanger
Cat	Ocelot	Tasmanian "tiger cat"
Wolf	Wolf	Tasmanian wolf

Parallel evolution of marsupial mammals in Australia and placental mammals on other continents.

Within the bats there has also been a remarkable adaptive radiation of food habits. Early in the history of the order, there evidently was a divergence into insectivorous (insect-eating) and frugivorous (fruit-eating) lines. The flying foxes (Megachiroptera) have generally maintained a fruit-eating habit, although some have become rather specialized nectar feeders. Members of the other major group (Microchiroptera) have been less conservative and have undergone considerable divergence in feeding habits. A majority of living microchiropterans are insectivorous, but members of two different families have become fish eaters. Within the large Neotropical family Phyllostomatidae, there are groups specialized to feed on fruit, nectar, insects, and small vertebrates (including other bats). Aberrant members of the family are the vampire bats, with a specialized dentition to aid blood lapping.

Form and Function

Skin and Hair

The skin of mammals is constructed of two layers, a superficial nonvascular epidermis and an inner layer, the dermis, or corium. The two layers interlock via fingerlike projections (dermal papillae), consisting of sensitive vascular dermis projecting into the epidermis. The outermost layers of the epidermis are cornified (impregnated with various tough proteins), and their cells are enucleate (lacking cell nuclei). The epidermis is composed of flattened cells in layers and is the interface between the individual and the environment. Its primary function is defensive, and it is cornified to resist abrasion. The surface of the skin is coated with lipids and organic salts, the so-called "acid mantle," which is thought to possess antifungal and antibacterial properties. Deep in the epidermis is an electronegative (electron-attracting) layer, a further deterrent to foreign organic or ionic agents.

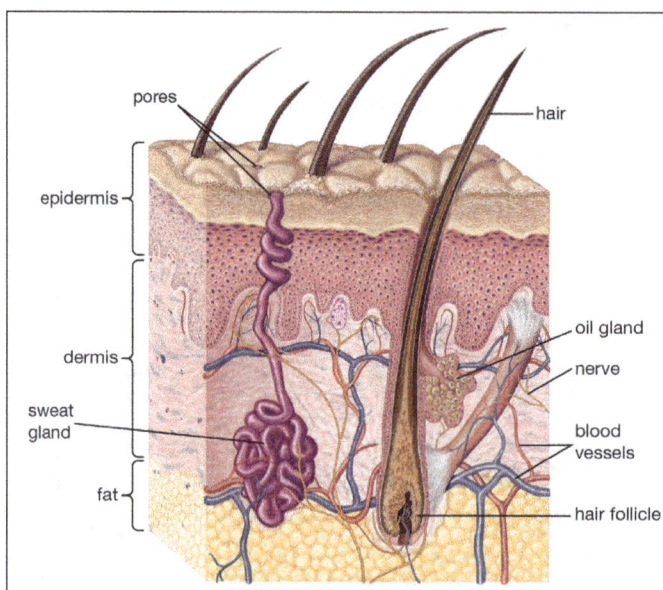

A cross section of mammalian skin and its underlying structures.

The dermis lies beneath the epidermis and nourishes it. The circulation of the dermis is variously developed in mammals, but it is typically extensive, out of proportion to the nutritional needs of the tissue. Its major role is to moderate body temperature and blood pressure by forming a peripheral shunt, an alternate route for the blood. Also in the dermis are sensory nerve endings to alert the individual to pressure (touch), heat, cold, and pain. In general, skin bearing hairs has few or no specialized sensory endings. The sensation of touch on hairy skin in humans depends on stimulation of the nerve fibres associated with the hairs. However, hairless skin, such as the lips and fingertips, has specialized endings.

Hair is derived from an invagination (pocketing) of the epidermis termed a follicle. Collectively, the hair is called the pelage. The individual hair is a rod of keratinized cells that may be cylindrical or more or less flattened. Keratin is a protein also found in claws and nails. The inner medulla of the hair is hollow and contains air; in the outer cortex layer there are frequently pigment granules. Associated with the hair follicle are nerve endings and a muscle, the arrector pili. The latter allows the erection of individual hairs to alter the insulative qualities of the pelage. The follicle also gives rise to sebaceous glands that produce sebum, a substance that lubricates the hair.

Most mammals have three distinct kinds of hairs. Guard hairs protect the rest of the pelage from abrasion and frequently from moisture, and they usually lend a characteristic colour pattern. The thicker underfur is primarily insulative and may differ in colour from the guard hairs. The third common hair type is the vibrissa, or whisker, a stiff, typically elongate hair that functions in tactile sensation. Hairs may be further modified to form rigid quills. The "horn" of the rhinoceros is composed of a fibrous keratin material derived from hair. Examples of keratinized derivatives of the integument other than hair are horns, hooves, nails, claws, and baleen.

Even though the primary function of the skin is defensive, it has been modified in mammals to serve such diverse functions as thermoregulation and nourishment of young. Secretions of sweat glands promote cooling due to evaporation at the surface of the body, and mammary glands are a type of apocrine gland (that is, a sweat gland associated with hair follicles).

In certain groups (primates in particular) the skin of the face is under intricate muscular control, and movements of the skin express and communicate emotion. In many mammals the colour and pattern of the pelage are important in communicative Behavior. Patterns may be startling (dymantic), as seen in the mane of the male lion or hamadryas baboon, warning (sematic), as seen in the bold pattern of skunks, or concealing (cryptic), perhaps the most common adaptation of pelage colour.

Hair has been secondarily lost or considerably reduced in some kinds of mammals. In adult cetaceans insulation is provided by thick subcutaneous fat deposits, or blubber, with hair limited to a few stiff vibrissae about the mouth. The bare skin is one of a number of features that contribute to the remarkably advanced hydrodynamics of locomotion in the group. Some burrowing (fossorial) mammals also tend toward reduction of the hair. This is shown most strikingly by the sand rats of northeastern Africa, but considerable loss of hair has also occurred in some species of pocket gophers. Hair may also be lost on restricted areas of the skin, as from the face in many monkeys or the buttocks of mandrills, and may be sparse on elephants and such highly modified species as pangolins and armadillos.

Continuous growth of hair (indeterminate), as seen on the heads of humans, is rare among mammals. Hairs with determinate growth are subject to wear and must be replaced periodically—a process termed molt. The first coat of a young mammal is referred to as the juvenal pelage, which typically is of fine texture like the underfur of adults and is replaced by a postjuvenile molt. Juvenal pelage is succeeded either directly by adult pelage or by the subadult pelage, which in some species is not markedly distinct from that of the adult. Once this pelage has been acquired, molting continues to recur at intervals, often annually or semiannually and sometimes more frequently. The pattern of molt typically is orderly, but it varies widely between species. Some mammals apparently molt continuously, with a few hairs at a time replaced throughout the year.

Teeth

Specialization in food habits has led to profound dental changes. The primitive mammalian tooth had high, sharp cusps and served to tear flesh or crush chitinous material (primarily the exoskeletons of terrestrial arthropods, such as insects). Herbivores tend to have specialized cheek teeth with complex patterns of contact (occlusion) and various ways of expanding the crowns of the teeth and circumventing the problem of wear. Omnivorous mammals, such as bears, pigs, and humans, tend to have molars with low, rounded cusps, termed bunodont.

A yawning African lion (Panthera leo) showing its long canine teeth.

A prime example of convergence in conjunction with dietary specialization is seen in those mammals adapted to feeding on ants and termites, a specialization generally termed myrmecophagy ("ant eating"). Trends frequently associated with myrmecophagy include strong claws, an elongate rounded skull, a wormlike extensible tongue, marked reduction in the mandible (lower jaw), and loss or extreme simplification of the teeth (dentition). This habit has led to remarkably similar morphology among animals as diverse as the echidna (a monotreme), the numbat (a marsupial), the anteater (a xenarthran), the aardvark (a tubulidentate), and the pangolin (a pholidotan).

Specialized herbivores evolved early in mammalian history. The extinct multituberculates were the earliest mammalian herbivores and have the longest evolutionary history, lasting more than 100 million years from 178 million to 50 million years ago. Multituberculate fossils, such as those of Ptilodus, dated to the Paleocene Epoch (66–56 million years ago) of North America, have been found on all continents. Similarities in teeth not due to common ancestry have occurred widely in herbivorous groups. Most herbivores have incisors modified for nipping or gnawing, have lost teeth with the resultant development of a gap (diastema) in the tooth row, and exhibit some molarization (expansion and flattening) of premolars to expand the grinding surface of the cheek teeth. Rootless incisors or cheek teeth have evolved frequently, their open pulp cavity allowing continual growth throughout life. Herbivorous specializations have evolved independently in multituberculates, rodents, lagomorphs, primates, and the wide diversity of ungulate and subungulate orders.

Skeleton

The mammalian skeletal system shows a number of advances over that of lower vertebrates. The mode of ossification (bone formation) of the long bones is characteristic. In lower vertebrates each long bone has a single centre of ossification (the diaphysis), and replacement of cartilage by bone proceeds from the centre toward the ends, which may remain cartilaginous, even in adults. In mammals secondary centres of ossification (the epiphyses) develop at the ends of the bones. Growth of bones occurs in zones of cartilage between diaphysis and epiphyses. Mammalian skeletal growth is termed determinate, for once the actively growing zone of cartilage has been obliterated, growth in length ceases. As in all bony vertebrates, of course, there is continual renewal of bone throughout life. The advantage of epiphyseal ossification lies in the fact that the bones have strong articular (joint-related) surfaces before the skeleton is mature. In general, the skeleton of the adult mammal has less structural cartilage than does that of a reptile.

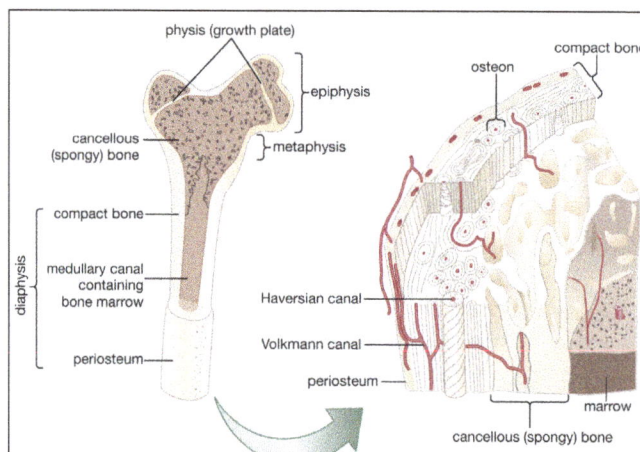

In above figure: Internal structure of a human long bone, with a magnified cross section of the interior. The central tubular region of the bone, called the diaphysis, flares outward near the end to form the metaphysis, which contains a largely cancellous, or spongy, interior. At the end of the bone is the epiphysis, which in young people is separated from the metaphysis by the physis, or growth plate. The periosteum is a connective sheath covering the outer surface of the bone. The Haversian system, consisting of inorganic substances arranged in concentric rings around the Haversian canals, provides compact bone with structural support and allows for metabolism of bone cells. Osteocytes (mature bone cells) are found in tiny cavities between the concentric rings. The canals contain capillaries that bring in oxygen and nutrients and remove wastes. Transverse branches are known as Volkmann canals.

The skeletal system of mammals and other vertebrates is broadly divisible functionally into axial and appendicular portions. The axial skeleton consists of the braincase (cranium) and the backbone and ribs, and it serves primarily to protect the central nervous system. The limbs and their girdles constitute the appendicular skeleton. In addition, there are skeletal elements derived from the gill arches of primitive vertebrates, collectively termed the visceral skeleton. Visceral elements in the mammalian skeleton include the jaws, the hyoid apparatus supporting the tongue, and the auditory ossicles of the middle ear. The postcranial axial skeleton in mammals generally has remained rather conservative during the course of evolution. The vast majority of mammals have seven cervical (neck) vertebrae; exceptions are sloths, with six or nine cervicals, and the sirenians with six. The anterior two cervical vertebrae are differentiated as atlas and axis. Specialized articulations of these two bones allow complex movements of the head on the trunk. Thoracic vertebrae bear ribs and are variable in number. The anterior ribs converge toward the ventral midline to articulate with the sternum, or breastbone, forming a semirigid thoracic "basket" for the protection of heart and lungs. Posterior to the thoracic region are the lumbar vertebrae, ranging from 2 to 21 in number (most frequently 4 to 7). Mammals have no lumbar ribs. There are usually 3 to 5 sacral vertebrae, but some xenarthrans have as many as 13. Sacral vertebrae fuse to form the sacrum, to which the pelvic girdle is attached. Caudal (tail) vertebrae range in number from 5 (fused elements of the human coccyx or tailbone) to 50.

The basic structure of the vertebral column is comparable throughout the Mammalia, although in many instances modifications have occurred in specialized locomotor modes to gain particular mechanical advantages. The vertebral column and associated muscles of many mammals are structurally analogous to a cantilever girder.

The skull is composite in origin and complex in function. Functionally the bones of the head are separable into the braincase and the jaws. In general, it is the head of the animal that meets the environment. The skull protects the brain and sense capsules (the parts of the skeleton that facilitate the senses of sight, hearing, taste, and smell), houses the teeth and tongue, and contains the entrance to the pharynx. Thus, the head functions in sensory reception, food acquisition, defense, respiration, and (in higher groups) communication. To serve these functions, bony elements have been recruited from the visceral skeleton, the endochondral skeleton (the parts of the skeleton that form from cartilage), and the dermal skeleton of lower vertebrates.

The skull of mammals differs markedly from that of reptiles because of the great expansion of the brain. The sphenoid bones that form the reptilian braincase form only the floor of the braincase in mammals. The side is formed in part by the alisphenoid bone, derived from the epipterygoid, a part of the reptilian palate. Dermal elements, the frontals and parietals, have come to lie deep to (beneath) the muscles of the jaw to form the dorsum of the braincase. Reptilian dermal roofing bones, lying superficial to the muscles of the jaw, are represented in mammals only by the jugal bone of the zygomatic arch, which lies under the eye.

In mammals a secondary palate is formed by processes of the maxillary bones and the palatines, with the pterygoid bones reduced in importance. The secondary palate separates the nasal passages from the oral cavity and allows continuous breathing while chewing or suckling.

Other specializations of the mammalian skull include paired articulating surfaces at the neck (occipital condyles) and an expanded nasal chamber with complexly folded turbinal bones, providing a large area for detection of odours. Eutherians have evolved bony protection for the middle ear, the auditory bulla. The development of this structure varies, although a ring-shaped (annular) tympanic bone is always present.

The bones of the mammalian middle ear are a diagnostic feature of the class. The three auditory ossicles form a series of levers that serve mechanically to increase the amplitude of sound waves reaching the tympanic membrane, or eardrum, produced as disturbances of the air. The innermost bone is the stapes, or "stirrup bone." It rests against the oval window of the inner ear. The stapes is homologous with the entire stapedial structure of reptiles, which in turn was derived from the hyomandibular arch of primitive vertebrates. The incus, or "anvil," articulates with the stapes. The incus was derived from the quadrate bone, which is involved in the jaw articulation in reptiles. The malleus, or "hammer," rests against the tympanic membrane and articulates with the incus. The malleus is the homologue of the reptilian articular bone. The mechanical efficiency of the middle ear has thus been increased by the incorporation of two bones of the reptilian jaw assemblage. In mammals the lower jaw is a single bone, the dentary, which articulates with the squamosal of the skull.

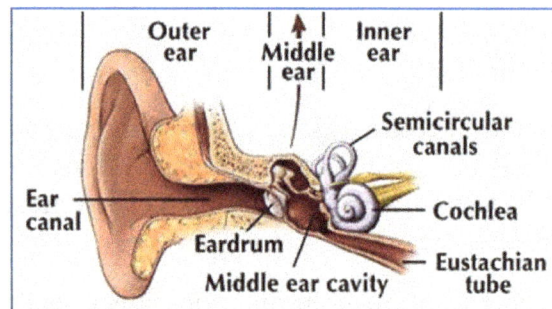

The structures of the outer, middle, and inner ear.

The limbs and girdles have been greatly modified with locomotor adaptations. The ancestral mammal had well-developed limbs and was five-toed. In each limb there were two distal (outer) elements (radius and ulna in the forelimb; tibia and fibula in the hind limb) and a single proximal (inner or upper) element (humerus; femur). There were nine bones in the wrist, the carpals, and seven bones in the ankle, the tarsals. The phalangeal formula (the number of phalangeal bones in each digit, numbered from inside outward) is 2-3-3-3-3 in primitive mammals; in primitive reptiles it is 2-3-4-5-3. Modifications in mammalian limbs have involved reduction, loss, or fusion of bones. Loss of the clavicle from the shoulder girdle, reduction in the number of toes, and modifications of tarsal and carpal bones are typical correlates of cursorial locomotion. Scansorial and arboreal groups tend to maintain or emphasize the primitive divergence of the thumb and hallux (the inner toe on the hind foot).

Centres of ossification sometimes develop in nonbony connective tissue. Such bones are termed heterotopic or sesamoid elements. The kneecap (patella) is such a bone. Another important bone of this sort, found in many kinds of mammals, is the baculum, or os penis, which occurs as a stiffening rod in the penis of such groups as carnivores, many bats, rodents, some insectivores, and many primates. The os clitoridis is a homologous structure found in females.

Muscles

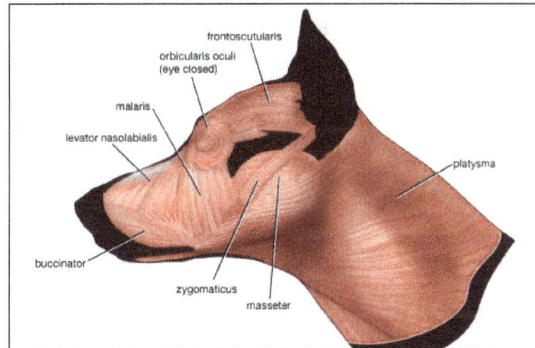

Facial musculature: Dog lateral view of facial musculature of a dog.

The muscular system of mammals is generally comparable to that of reptiles. With changes in locomotion, the proportions and specific functions of muscular elements have been altered, but the relationships of these muscles remain essentially the same. Exceptions to this generalization are the muscles of the skin and of the jaw.

The panniculus carnosus is a sheath of dermal (skin) muscle, developed in many mammals, that allows the movement of the skin independent of the movement of deeper muscle masses. These movements function in such mundane activities as the twitching of the skin to foil insect pests and in some species also are important in shivering, a characteristic heat-producing response to thermal stress. The dermal musculature of the facial region is particularly well developed in primates and carnivores but occurs in other groups as well. Facial mobility allows expression that may be of importance in the behavioral maintenance of interspecific social structure.

The temporalis muscle is the major adductor (closer) of the reptilian jaw. In mammals the temporalis is divided into a deep temporalis proper and a more superficial masseter muscle. The temporalis attaches to the coronoid process of the mandible (lower jaw) and the temporal bone of the skull. The masseter passes from the angular process of the mandible to the zygomatic arch. The masseter allows an anteroposterior (forward-backward) movement of the jaw and is highly developed in mammals, such as rodents, for which grinding is the important function of the dentition.

Digestive System

The alimentary canal is highly specialized in many kinds of mammals. In general, specializations of the gut accompany herbivorous habits. The intestines of herbivores are typically elongate, and the stomach may also be specialized. Subdivision of the gut allows areas of differing physiological environments for the activities of different sorts of enzymes and symbiotic bacteria, which aid the animal by breaking down certain compounds that are otherwise undigestible. In ruminant artiodactyls, such as antelopes, deer, and cattle, the stomach has up to four chambers, each with a particular function in the processing of vegetable material. A cecum is common in many herbivores. The cecum is a blind sac at the far end of the small intestine where complex compounds such as cellulose are acted upon by symbiotic bacteria. The vermiform appendix is a diverticulum of the cecum. The appendix is rich in lymphoid tissue and in many mammals is concerned with defense against toxic bacterial products.

Hares and rabbits, the sewellel, or "mountain beaver" (Aplodontia rufa), and some insectivores exhibit a phenomenon of reingestion called coprophagy, in which at intervals specialized fecal pellets are produced. These pellets are eaten and passed through the alimentary canal a second time. Where known to be present, this pattern seems to be obligatory. Reingestion primarily occurs in members of the shrew, rodent, and rabbit groups; however, the Behavior has been observed to a lesser degree in other groups, including canines and pikas. The process appears to allow the animal to absorb in the upper gut vitamins produced by the microflora of the lower gut but not absorbable there.

Excretory System

The mammalian kidney is constructed of a large number of functional units called nephrons. Each nephron consists of a distal tubule, a medial section termed the loop of Henle, a proximal tubule, and a renal corpuscle. The renal corpuscle is a knot of capillaries (glomerulus) surrounded by a sheath (Bowman's capsule). The renal corpuscle is a pressure filter, relying on blood pressure to remove water, ions, and small organic molecules from the blood. Some of the material removed is waste, but some is of value to the organism. The filtrate is sorted by the tubules, and water and needed solutes are resorbed. Resorption is both passive (osmotic) and active (based on ion transport systems). The distal convoluted tubules drain into collecting tubules, which in turn empty into the calyces, or branches, of the renal pelvis, the expanded end of the ureter. The pressure-pump nephron of mammals is so efficient that the renal portal system of lower vertebrates has been completely lost. Mammalian kidneys show considerable variety in structure, relative to the environmental demands on a given species. In particular, desert rodents have long loops of Henle and are able to resorb much water and to excrete a highly concentrated urine. Urea is the end product of protein metabolism in mammals, and excretion is therefore called ureotelic.

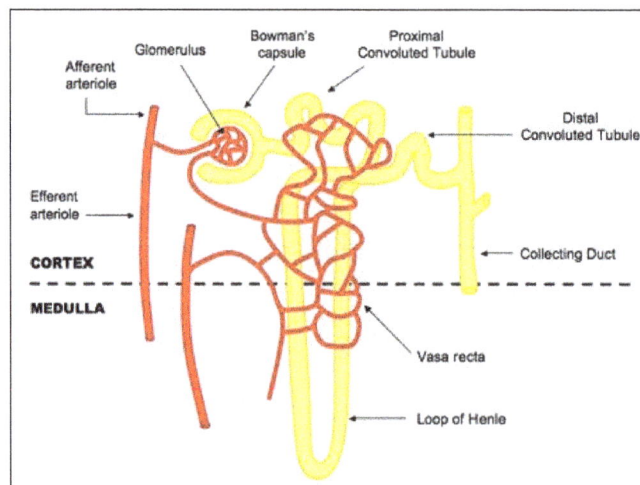

Each nephron of the kidney contains blood vessels and a special tubule. As the filtrate flows through the tubule of the nephron, it becomes increasingly concentrated into urine. Waste products are transferred from the blood into the filtrate, while nutrients are absorbed from the filtrate into the blood.

Reproductive System

The Male Tract

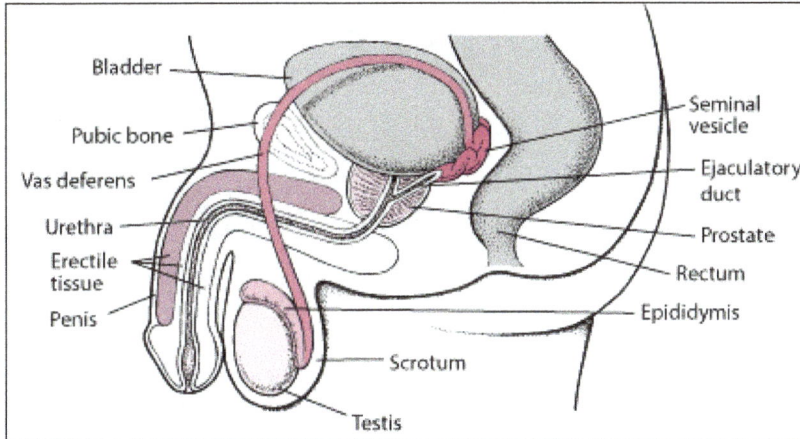

Sagittal section of the male reproductive organs,
showing the prostate gland and seminal vesicles.

The testes of mammals descend from the abdominal cavity to lie in a compartmented pouch termed the scrotum. In some species the testes are permanently scrotal, and the scrotum is sealed off from the general abdominal cavity. In other species the testes migrate to the scrotum only during the breeding season. It is thought that the temperature of the abdominal cavity is too high to allow spermatogenesis; the scrotum allows cooling of the testes.

The transport of spermatozoa is comparable to that in reptiles, relying on ducts derived from urinary ducts of earlier vertebrates. Mammalian specialities are the bulbourethral (or Cowper's) glands, the prostate gland, and the seminal vesicle or vesicular gland. Each of these glands adds secretions to the spermatozoa to form semen, which passes from the body via a canal (urethra) in the highly vascular, erectile penis. The tip of the penis, the glans, may have a complex morphology and has been used as a taxonomic character in some groups. The penis may be retracted into a sheath along the abdomen or may be pendulous, as in bats and many primates.

The Female Tract

The structure of the female reproductive tract is variable. Four types of uterus are generally recognized among placentals, based on the relationship of the uterine horns (branches). A duplex uterus characterizes rodents and rabbits; the uterine horns are completely separated and have separate cervices opening into the vagina. Carnivores have a bipartite uterus, in which the horns are largely separate but enter the vagina by a single cervix. In the bicornate uterus, typical of many ungulates, the horns are distinct for less than half their length; the lower part of the uterus is a common chamber, the body. Higher primates have a simplex uterus in which there is no separation between the horns and thus a single chamber.

The female reproductive tract of marsupials is termed didelphous; the vagina is paired, as are oviducts and uteri. In primitive marsupials there are paired vaginae lateral to the ureters. In more advanced groups, such as kangaroos, the lateral vaginae persist and conduct the migration of spermatozoa, but a medial "pseudovagina" functions as the birth canal.

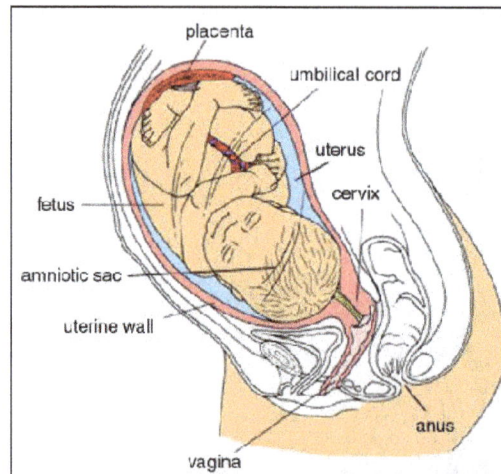

The uterus is an inverted pear-shaped muscular organ of the female reproductive
system, located between the bladder and the rectum. It functions to nourish
and house the fertilized egg until the unborn child is ready to be delivered.

Monotremes have paired uteri and oviducts, which empty into a urogenital sinus (cavity) as fluid wastes
do. The sinus passes into the cloaca, a common receptacle for reproductive and excretory products.

Circulatory System

In mammals, as in birds, the right and left ventricles of the heart are completely separated, so
that pulmonary (lung) and systemic (body) circulations are completely independent. Oxygenated
blood arrives in the left atrium from the lungs and passes to the left ventricle, whence it is forced
through the aorta to the systemic circulation. Deoxygenated blood from the tissues returns to
the right atrium via a large vein, the vena cava, and is pumped to the pulmonary capillary bed
through the pulmonary artery.

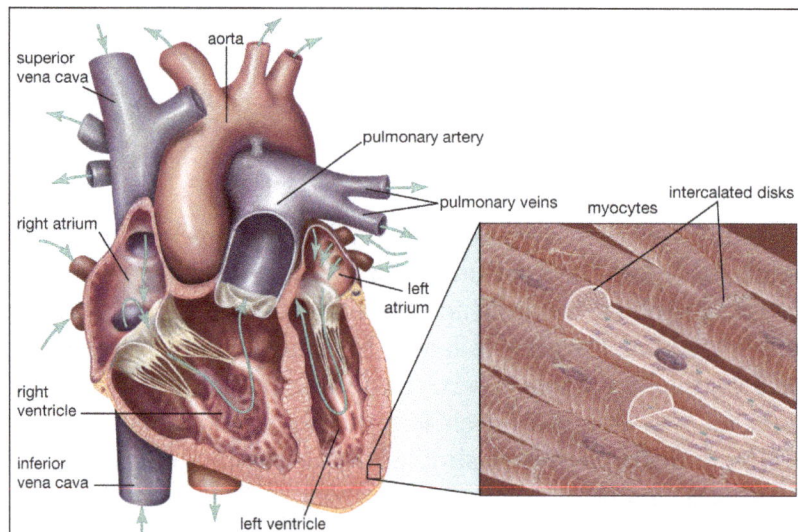

Cross section of a four-chambered mammalian heart.

Among vertebrates contraction of the heart is myogenic, or generated by muscle; rhythm is in-
herent in all cardiac muscle, but in myogenic hearts the pacemaker is derived from cardiac tissue.

The pacemaker in mammals (and also in birds) is an oblong mass of specialized cells called the sinoatrial node, located in the right atrium near the junction with the venae cavae. A wave of excitation spreads from this node to the atrioventricular node, which is located in the right atrium near the base of the interatrial septum. From this point excitation is conducted along the atrioventricular bundle (bundle of His) and enters the main mass of cardiac tissue along fine branches, the Purkinje fibres. Homeostatic, or stable, control of the heart by neuroendocrine or other agents is mediated through the intrinsic control network of the heart.

Blood leaves the left ventricle through the aorta. The mammalian aorta is an unpaired structure derived from the left fourth aortic arch of the primitive vertebrate. Birds, on the other hand, retain the right fourth arch.

The circulatory system forms a complex communication and distribution network to all physiologically active tissues of the body. A constant, copious supply of oxygen is required for sustaining the active, heat-producing (endothermous) physiology of the higher vertebrates. The efficiency of the four-chambered heart is important to this function. Oxygen is transported by specialized red blood cells, or erythrocytes, as in all vertebrates. Packaging the oxygen-bearing pigment hemoglobin in erythrocytes keeps the viscosity of the blood minimal and thereby allows efficient circulation while limiting the mechanical load on the heart. The mammalian erythrocyte is a highly evolved structure; its discoid, biconcave shape allows maximal surface area per unit volume. When mature and functional, mammalian red blood cells are enucleate (lacking a nucleus).

Respiratory System

Closely coupled with the circulatory system is the ventilatory (breathing) apparatus, the lungs and associated structures. Ventilation in mammals is unique. The lungs themselves are less efficient than those of birds, for air movement consists of an ebb and flow, rather than a one-way circuit, so a residual volume of air always remains that cannot be expired. Ventilation in mammals is by means of a negative pressure pump made possible by the evolution of a definitive thoracic cavity with a diaphragm.

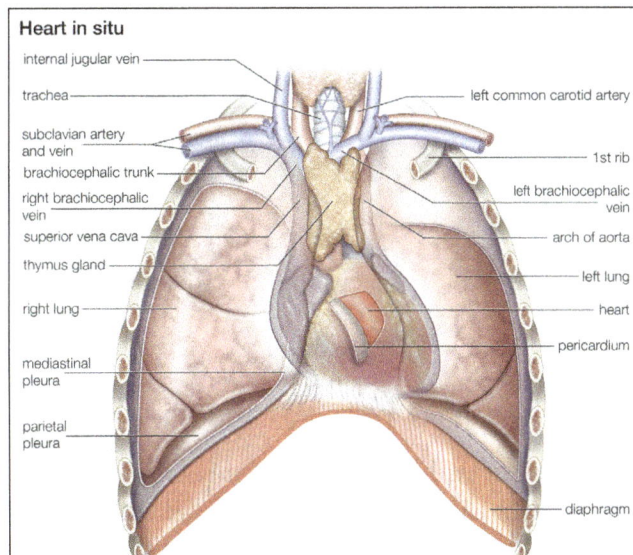

Heart in situ

internal jugular vein —

trachea —

subclavian artery and vein

brachiocephalic trunk —

right brachiocephalic vein —

superior vena cava —

thymus gland —

right lung —

mediastinal pleura —

parietal pleura —

— left common carotid artery

— 1st rib

— left brachiocephalic vein

— arch of aorta

— left lung

— heart

— pericardium

— diaphragm

The human heart in situ.

The diaphragm is a unique composite structure consisting of (1) the transverse septum (a wall that primitively separates the heart from the general viscera); (2) pleuroperitoneal folds from the body wall; (3) mesenteric folds; and (4) axial muscles inserting on a central tendon, or diaphragmatic aponeurosis.

The lungs lie in separate airtight compartments called pleural cavities, separated by the mediastinum. As the size of the pleural cavity is increased, the lung is expanded and air flows in passively. Enlargement of the pleural cavity is produced by contraction of the diaphragm or by elevation of the ribs. The relaxed diaphragm domes upward, but when contracted it stretches flat. Expiration is an active movement brought about by contraction of abdominal muscles against the viscera.

Air typically enters the respiratory passages through the nostrils, where it may be warmed and moistened. It passes above the bony palate and the soft palate and enters the pharynx. In the pharynx the passages for air and food cross. Air enters the trachea, which divides at the level of the lungs into primary bronchi. A characteristic feature of the trachea of many mammals is the larynx. Vocal cords stretch across the larynx and are vibrated by forced expiration to produce sound. The laryngeal apparatus may be greatly modified for the production of complex vocalizations. In some groups—for example, howler monkeys—the hyoid apparatus is incorporated into the sound-producing organ, as a resonating chamber.

Nervous and Endocrine Systems

The nervous system and the endocrine system are closely related to one another in their function, for both serve to coordinate activity. The endocrine glands of mammals generally have more complex regulatory functions than do those of lower vertebrates. This is particularly true of the pituitary gland, which supplies hormones that regulate the reproductive cycle. Follicle-stimulating hormone (FSH) initiates the maturation of the ovarian follicle. Luteinizing hormone (LH) mediates the formation of the corpus luteum from the follicle following ovulation. Prolactin, also a product of the anterior pituitary, stimulates the secretion of milk.

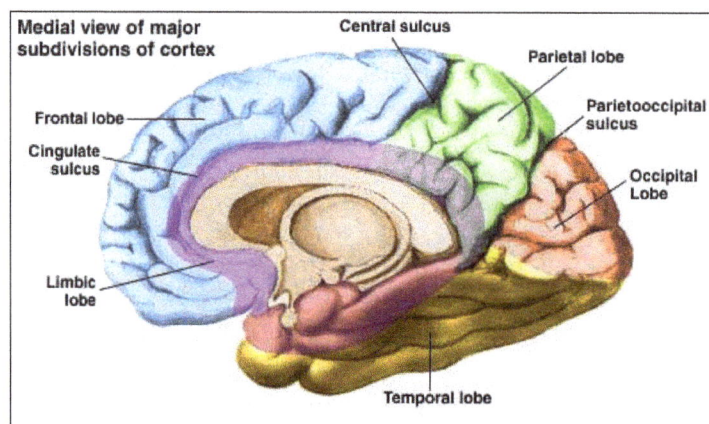

Medial view of the left hemisphere of the human brain.

Control of the pituitary glands is partially by means of neurohumours from the hypothalamus, a part of the forebrain in contact with the pituitary gland by nervous and circulatory pathways. The hypothalamus is of the utmost importance in mammals, for it integrates stimuli from both internal and external environments, channeling signals to higher centres or into autonomic pathways.

The cerebellum of vertebrates is at the anterior end of the hindbrain. Its function is to coordinate motor activities and to maintain posture. In most mammals the cerebellum is highly developed, and its surface may be convoluted to increase its area. The data with which the cerebellum works arrive from proprioceptors ("self-sensors") in the muscles and from the membranous labyrinth of the inner ear, the latter giving information on position and movements of the head.

In the vertebrate ancestors of mammals, the cerebral hemispheres were centres for the reception of olfactory stimuli. Vertebrate evolution has favoured an increasing importance of these lobes in the integration of stimuli. Their great development in mammals as centres of association is responsible for the "creative" Behavior of members of the class—i.e., the ability to learn, to adapt as individuals to short-term environmental change through appropriate responses on the basis of previous experience. In vertebrate evolution the gray matter of the cerebrum has moved from a primitive internal position in the hemispheres to a superficial position. The superficial gray matter is termed the pallium. The paleopallium of amphibians has become the olfactory lobes of the higher vertebrates; the dorsolateral surface, or archipallium, has become the mammalian hippocampus. The great neural advance of the mammals lies in the elaboration of the neopallium, which makes up the bulk of the cerebrum. The neopallium is an association centre, the dominant centre of neural function, and is involved in so-called "intelligent" response. By contrast, the highest centre in the avian brain is the corpus striatum, an evolutionary product of the basal nuclei of the amphibian brain. Therefore, the bulk of complex Behavior of birds is instinctive. The surface of the neopallium tends in some mammals to be greatly expanded by convoluting, forming folds (gyri) between deep grooves (sulci).

Classification

The higher classification of the class Mammalia is based on consideration of a broad array of characters. Traditionally, evidence from comparative anatomy was of predominant importance, but, more recently, information from such disciplines as physiology, serology (the study of immune reactions in body fluids), and genetics has proved useful in considering relationships. Comparative study of living organisms is supplemented by the findings of paleontology. Study of the fossil record adds a historical dimension to knowledge of mammalian relationships. In some cases—the horses, for example—the fossil record has been adequate to allow lineages to be traced in great detail.

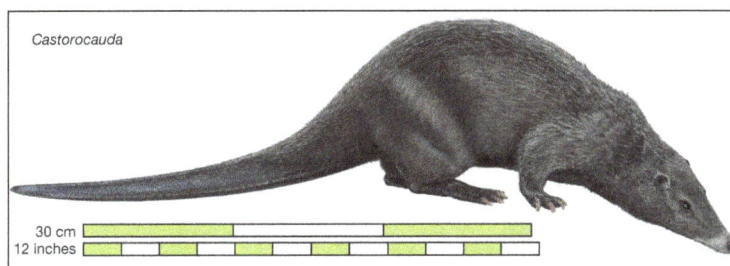

Weighed 500 to 800 grams (1.1 to 1.8 pounds) and
was almost as large as living platypuses, making it the largest Jurassic mammal known.

Relative to that of other major vertebrate groups, the fossil record of mammals is good. Fossilization depends upon a great many factors, the most important of which are the structure of the organism, its habitat, and conditions at the time of death. The most common remains of mammals are teeth and the associated bones of the jaw and skull. Enamel covering the typical mammalian tooth is composed of prismatic rods of crystalline apatite and is the hardest tissue in the mammalian body. It is highly resistant to chemical and physical weathering. Because of the abundance of teeth in deposits of fossil mammals, dental characteristics have been stressed in the interpretation of mammalian phylogeny and relationships. Dental features are particularly well suited for this important role in classification because they reflect the broad radiation of mammalian feeding specializations from the primitive predaceous habit.

This classification is modified from that of Malcolm C. McKenna and Susan K. Bell to classify higher categories of mammals with significant contributions from Don E. Wilson and DeeAnn M. Reeder; extinct groups are not listed.

- Class mammalia (mammals):

  Approximately 5,500 species in 29 orders.

- Subclass Prototheria (monotremes, egg-laying mammals):

  Here 5 species are classified in 2 orders, but monotremes have traditionally been classified together in a single order, Monotremata.

- Order Tachyglossa (echidnas):

  4 species in 1 family.

- Order Platypoda (platypus):

  1 species.

- Subclass Theria (live-bearing mammals).

- Metatheria (marsupials):

  More than 330 species in 7 orders.

- Order Diprotodontia (kangaroos, koalas, wombats, possums, and kin):

  More than 140 species in 11 families.

- Order Dasyuromorphia (carnivorous marsupials):

  About 70 species in 3 families, not including the extinct thylacine (Tasmanian wolf), sole member of family Thylacinidae.

- Order Peramelemorphia (bandicoots and bilbies):

  About 21 species in 3 families.

- Order Notoryctemorphia (marsupial moles):

  2 species in 1 family.

- Order Microbiotheria (monito del monte):

  1 species (Dromiciops gliroides).

- Order Didelphimorphia (opossums):

  About 90 species in 1 family.

- Order Paucituberculata (shrew, or rat, opossums):

  6 species in 1 family.

- Eutheria (placental mammals):

  More than 5,000 species in 20 orders.

- Order Rodentia (rodents):

  Nearly 2,300 species in 30 families.

- Order Chiroptera (bats):

  More than 1,100 species in 18 families.

- Order Soricomorpha (shrews, moles, and kin):

  About 430 species in 4 families. Moles (family Talpidae) are sometimes classified with hedgehogs in Erinaceomorpha.

- Order Afrosoricida (golden moles and tenrecs):

  About 50 species in 2 families.

- Order Erinaceomorpha (hedgehogs):

  24 species in 1 family.

- Order Primates (humans, apes, monkeys, lemurs, and kin):

  About 375 species in 15 families. Colugos are sometimes classified as a separate order, Dermoptera.

- Grandorder Ungulata (ungulates):

  About 350 species in 7 orders.

- Order Artiodactyla (even-toed hoofed ungulates):

  About 240 species in 10 families, including giraffes, camels, deer, cattle, pigs, sheep, goats, and kin.

- Order Cetacea (whales, dolphins, and porpoises):

  More than 80 species in 11 families.

- Order Perissodactyla (odd-toed hoofed ungulates).

- 17 species in 3 families, including horses, rhinoceroses, tapirs, and kin.

- Uranotherians:

  The following three ungulate orders (Sirenia, Proboscidea, and Hyracoidea) are sometimes grouped together as the order Uranotheria, for they are more closely related to one another than to other ungulates.

- Order Hyracoidea (hyraxes):

  4 species in 1 family.

- Order Sirenia (manatees and dugongs):

  5 species in 2 families.

- Order Proboscidea (elephants):

  3 species in 1 family.

- Order Tubulidentata (aardvark):

  1 species (Orycteropus afer).

- Order Carnivora (carnivores):

  Nearly 290 species in 15 families.

- Order Lagomorpha (pikas and rabbits):

  92 species in 3 families.

- Magnorder Xenarthra (edentates, or xenarthrans):

  31 species in 2 orders.

- Order Cingulata (armadillos):

  20 species in 1 family.

- Order Pilosa (anteaters and sloths):

  10 species in 4 families.

- Order Scandentia (tree shrews):

  20 species in 2 families.

- Order Macroscelidea (elephant shrews):

  15 species in 1 family.

- Order Pholidota (pangolins):

  8 species in 1 family.

FISH

Fish are gill-bearing aquatic craniate animals that lack limbs with digits. They form a sister group to the tunicates, together forming the olfactores. Included in this definition are the living hagfish, lampreys, and cartilaginous and bony fish as well as various extinct related groups.

Tetrapods emerged within lobe-finned fishes, so cladistically they are fish as well. However, traditionally fish are rendered paraphyletic by excluding the tetrapods (i.e., the amphibians, reptiles, birds and mammals which all descended from within the same ancestry). Because in this manner the term "fish" is defined negatively as a paraphyletic group, it is not considered a formal taxonomic grouping in systematic biology, unless it is used in the cladistic sense, including tetrapods. The traditional term pisces (also ichthyes) is considered a typological, but not a phylogenetic classification.

The earliest organisms that can be classified as fish were soft-bodied chordates that first appeared during the Cambrian period. Although they lacked a true spine, they possessed notochords which allowed them to be more agile than their invertebrate counterparts. Fish would continue to evolve through the Paleozoic era, diversifying into a wide variety of forms. Many fish of the Paleozoic developed external armor that protected them from predators. The first fish with jaws appeared in the Silurian period, after which many (such as sharks) became formidable marine predators rather than just the prey of arthropods.

Most fish are ectothermic ("cold-blooded"), allowing their body temperatures to vary as ambient temperatures change, though some of the large active swimmers like white shark and tuna can hold a higher core temperature.

Fish can communicate in their underwater environments through the use of acoustic communication. Acoustic communication in fish involves the transmission of acoustic signals from one individual of a species to another. The production of sounds as a means of communication among fish is most often used in the context of feeding, aggression or courtship Behavior. The sounds emitted by fish can vary depending on the species and stimulus involved. They can produce either stridulatory sounds by moving components of the skeletal system, or can produce non-stridulatory sounds by manipulating specialized organs such as the swimbladder.

 Fish are abundant in most bodies of water. They can be found in nearly all aquatic environments, from high mountain streams (e.g., char and gudgeon) to the abyssal and even hadal depths of the deepest oceans (e.g., gulpers and anglerfish), although no species has yet been documented in the deepest 25% of the ocean. With 33,600 described species, fish exhibit greater species diversity than any other group of vertebrates.

Fish are an important resource for humans worldwide, especially as food. Commercial and subsistence fishers hunt fish in wild fisheries or farm them in ponds or in cages in the ocean. They are also caught by recreational fishers, kept as pets, raised by fishkeepers, and exhibited in public aquaria. Fish have had a role in culture through the ages, serving as deities, religious symbols, and as the subjects of art, books and movies.

Anatomy and Physiology

Gills

Most fish exchange gases using gills on either side of the pharynx. Gills consist of threadlike structures called filaments. Each filament contains a capillary network that provides a large surface area for exchanging oxygen and carbon dioxide. Fish exchange gases by pulling oxygen-rich water through their mouths and pumping it over their gills. In some fish, capillary blood flows in the opposite direction to the water, causing countercurrent exchange. The gills push the oxygen-poor water out through openings in the sides of the pharynx. Some fish, like sharks and lampreys, possess multiple gill openings. However, bony fish have a single gill opening on each side. This opening is hidden beneath a protective bony cover called an operculum.

Juvenile bichirs have external gills, a very primitive feature that they share with larval amphibians.

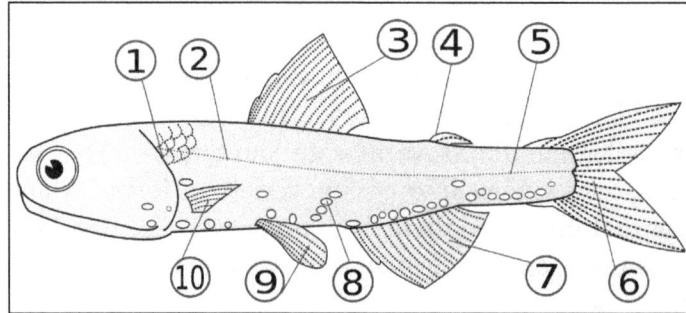

The anatomy of Lampanyctodes hectoris (1) – operculum (gill cover), (2) – lateral line, (3) – dorsal fin, (4) – fat fin, (5) – caudal peduncle, (6) – caudal fin, (7) – anal fin, (8) – photophores, (9) – pelvic fins (paired), (10) – pectoral fins (paired).

Air Breathing

Fish from multiple groups can live out of the water for extended periods. Amphibious fish such as the mudskipper can live and move about on land for up to several days,[dubious – discuss] or live in stagnant or otherwise oxygen depleted water. Many such fish can breathe air via a variety of mechanisms. The skin of anguillid eels may absorb oxygen directly. The buccal cavity of the electric eel may breathe air. Catfish of the families Loricariidae, Callichthyidae, and Scoloplacidae absorb air through their digestive tracts. Lungfish, with the exception of the Australian lungfish, and bichirs have paired lungs similar to those of tetrapods and must surface to gulp fresh air through the mouth and pass spent air out through the gills. Gar and bowfin have a vascularized swim bladder that functions in the same way. Loaches, trahiras, and many catfish breathe by passing air through the gut. Mudskippers breathe by absorbing oxygen across the skin (similar to frogs). A number of fish have evolved so-called accessory breathing organs that extract oxygen from the air. Labyrinth fish (such as gouramis and bettas) have a labyrinth organ above the gills that performs this function. A few other fish have structures resembling labyrinth organs in form and function, most notably snakeheads, pikeheads, and the Clariidae catfish family.

Breathing air is primarily of use to fish that inhabit shallow, seasonally variable waters where the water's oxygen concentration may seasonally decline. Fish dependent solely on dissolved oxygen, such as perch and cichlids, quickly suffocate, while air-breathers survive for much longer, in some cases in water that is little more than wet mud. At the most extreme, some air-breathing fish are able to survive in damp burrows for weeks without water, entering a state of aestivation (summertime hibernation) until water returns.

Air breathing fish can be divided into obligate air breathers and facultative air breathers. Obligate air breathers, such as the African lungfish, must breathe air periodically or they suffocate. Facultative air breathers, such as the catfish Hypostomus plecostomus, only breathe air if they need to and will otherwise rely on their gills for oxygen. Most air breathing fish are facultative air breathers that avoid the energetic cost of rising to the surface and the fitness cost of exposure to surface predators.

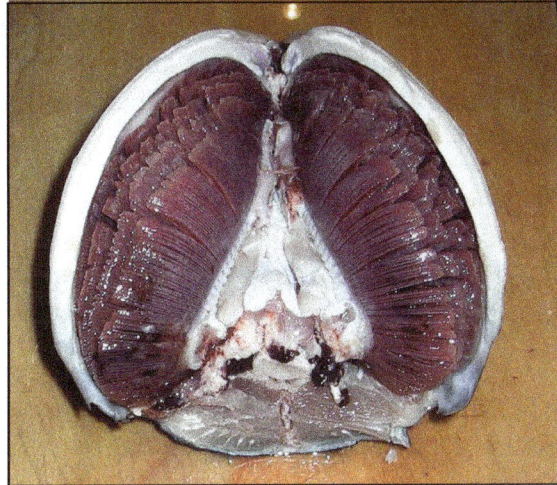

Tuna gills inside the head. The fish head is oriented snout-downwards,
with the view looking towards the mouth.

Circulation

Fish have a closed-loop circulatory system. The heart pumps the blood in a single loop throughout the body. In most fish, the heart consists of four parts, including two chambers and an entrance and exit. The first part is the sinus venosus, a thin-walled sac that collects blood from the fish's veins before allowing it to flow to the second part, the atrium, which is a large muscular chamber. The atrium serves as a one-way antechamber, sends blood to the third part, ventricle. The ventricle is another thick-walled, muscular chamber and it pumps the blood, first to the fourth part, bulbus arteriosus, a large tube, and then out of the heart. The bulbus arteriosus connects to the aorta, through which blood flows to the gills for oxygenation.

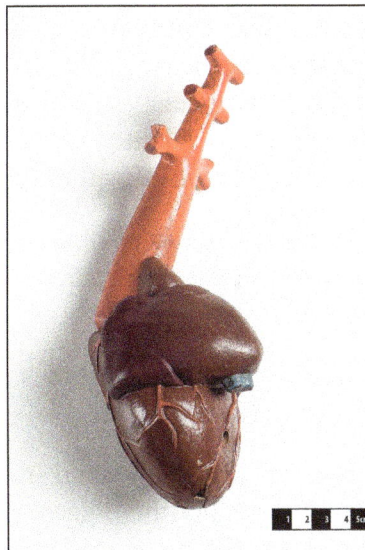

Didactic model of a fish heart.

Digestion

Jaws allow fish to eat a wide variety of food, including plants and other organisms. Fish ingest food through the mouth and break it down in the esophagus. In the stomach, food is further digested and, in many fish, processed in finger-shaped pouches called pyloric caeca, which secrete digestive enzymes and absorb nutrients. Organs such as the liver and pancreas add enzymes and various chemicals as the food moves through the digestive tract. The intestine completes the process of digestion and nutrient absorption.

Excretion

As with many aquatic animals, most fish release their nitrogenous wastes as ammonia. Some of the wastes diffuse through the gills. Blood wastes are filtered by the kidneys.

Saltwater fish tend to lose water because of osmosis. Their kidneys return water to the body. The reverse happens in freshwater fish: they tend to gain water osmotically. Their kidneys produce dilute urine for excretion. Some fish have specially adapted kidneys that vary in function, allowing them to move from freshwater to saltwater.

Scales

The scales of fish originate from the mesoderm (skin); they may be similar in structure to teeth.

Sensory and Nervous System

Central Nervous System

Fish typically have quite small brains relative to body size compared with other vertebrates, typically one-fifteenth the brain mass of a similarly sized bird or mammal. However, some fish have relatively large brains, most notably mormyrids and sharks, which have brains about as massive relative to body weight as birds and marsupials.

Fish brains are divided into several regions. At the front are the olfactory lobes, a pair of structures that receive and process signals from the nostrils via the two olfactory nerves. The olfactory lobes are very large in fish that hunt primarily by smell, such as hagfish, sharks, and catfish. Behind the olfactory lobes is the two-lobed telencephalon, the structural equivalent to the cerebrum in higher vertebrates. In fish the telencephalon is concerned mostly with olfaction. Together these structures form the forebrain.

Connecting the forebrain to the midbrain is the diencephalon. The diencephalon performs functions associated with hormones and homeostasis. The pineal body lies just above the diencephalon. This structure detects light, maintains circadian rhythms, and controls color changes.

The midbrain (or mesencephalon) contains the two optic lobes. These are very large in species that hunt by sight, such as rainbow trout and cichlids.

The hindbrain (or metencephalon) is particularly involved in swimming and balance. The cerebellum is a single-lobed structure that is typically the biggest part of the brain. Hagfish and lampreys have relatively small cerebellae, while the mormyrid cerebellum is massive and apparently involved in their electrical sense.

The brain stem (or myelencephalon) is the brain's posterior. As well as controlling some muscles and body organs, in bony fish at least, the brain stem governs respiration and osmoregulation.

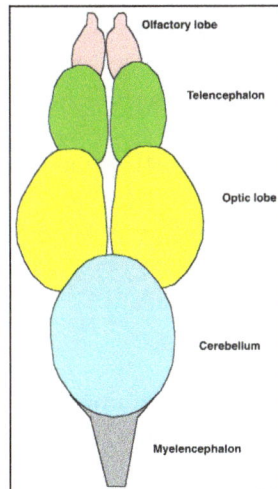

Dorsal view of the brain of the rainbow trout.

Sense Organs

Most fish possess highly developed sense organs. Nearly all daylight fish have color vision that is at least as good as a human's. Many fish also have chemoreceptors that are responsible for extraordinary senses of taste and smell. Although they have ears, many fish may not hear very well. Most fish have sensitive receptors that form the lateral line system, which detects gentle currents and vibrations, and senses the motion of nearby fish and prey. Some fish, such as catfish and sharks, have the Ampullae of Lorenzini, organs that detect weak electric currents on the order of millivolt. Other fish, like the South American electric fishes Gymnotiformes, can produce weak electric currents, which they use in navigation and social communication.

Fish orient themselves using landmarks and may use mental maps based on multiple landmarks or symbols. Fish behavior in mazes reveals that they possess spatial memory and visual discrimination.

Vision

Vision is an important sensory system for most species of fish. Fish eyes are similar to those of terrestrial vertebrates like birds and mammals, but have a more spherical lens. Their retinas generally have both rods and cones (for scotopic and photopic vision), and most species have colour vision. Some fish can see ultraviolet and some can see polarized light. Amongst jawless fish, the lamprey has well-developed eyes, while the hagfish has only primitive eyespots. Fish vision shows adaptation to their visual environment, for example deep sea fishes have eyes suited to the dark environment.

Hearing

Hearing is an important sensory system for most species of fish. Fish sense sound using their lateral lines and their ears.

Cognition

New research has expanded preconceptions about the cognitive capacities of fish. For example, manta rays have exhibited behavior linked to self-awareness in mirror test cases. Placed in front of a mirror, individual rays engaged in contingency testing, that is, repetitive behavior aiming to check whether their reflection's behavior mimics their body movement.

Wrasses have also passed the mirror test in a 2018 scientific study.

Cases of tool use have also been noticed, notably in the Choerodon family, in archerfish and Atlantic cod.

Capacity for Pain

Experiments done by William Tavolga provide evidence that fish have pain and fear responses. For instance, in Tavolga's experiments, toadfish grunted when electrically shocked and over time they came to grunt at the mere sight of an electrode.

In 2003, Scottish scientists at the University of Edinburgh and the Roslin Institute concluded that rainbow trout exhibit behaviors often associated with pain in other animals. Bee venom and acetic acid injected into the lips resulted in fish rocking their bodies and rubbing their lips along the sides and floors of their tanks, which the researchers concluded were attempts to relieve pain, similar to what mammals would do. Neurons fired in a pattern resembling human neuronal patterns.

Professor James D. Rose of the University of Wyoming claimed the study was flawed since it did not provide proof that fish possess "conscious awareness, particularly a kind of awareness that is meaningfully like ours". Rose argues that since fish brains are so different from human brains, fish are probably not conscious in the manner humans are, so that reactions similar to human reactions to pain instead have other causes. Rose had published a study a year earlier arguing that fish cannot feel pain because their brains lack a neocortex. However, animal behaviorist Temple Grandin argues that fish could still have consciousness without a neocortex because "different species can use different brain structures and systems to handle the same functions."

Animal welfare advocates raise concerns about the possible suffering of fish caused by angling. Some countries, such as Germany have banned specific types of fishing, and the British RSPCA now formally prosecutes individuals who are cruel to fish.

Muscular System

Swim bladder of a rudd (Scardinius erythrophthalmus).

Most fish move by alternately contracting paired sets of muscles on either side of the backbone. These contractions form S-shaped curves that move down the body. As each curve reaches the back fin, backward force is applied to the water, and in conjunction with the fins, moves the fish forward. The fish's fins function like an airplane's flaps. Fins also increase the tail's surface area, increasing speed. The streamlined body of the fish decreases the amount of friction from the water. Since body tissue is denser than water, fish must compensate for the difference or they will sink. Many bony fish have an internal organ called a swim bladder that adjusts their buoyancy through manipulation of gases.

Endothermy

Although most fish are exclusively ectothermic, there are exceptions. The only known bony fishes (infraclass Teleostei) that exhibit endothermy are in the suborder Scombroidei – which includes the billfishes, tunas, and the butterfly kingfish, a basal species of mackerel – and also the opah. The opah, a lampriform, was demonstrated in 2015 to utilize "whole-body endothermy", generating heat with its swimming muscles to warm its body while countercurrent exchange (as in respiration) minimizes heat loss. It is able to actively hunt prey such as squid and swim for long distances due to the ability to warm its entire body, including its heart, which is a trait typically found in only mammals and birds (in the form of homeothermy). In the cartilaginous fishes (class Chondrichthyes), sharks of the families Lamnidae (porbeagle, mackerel, salmon, and great white sharks) and Alopiidae (thresher sharks) exhibit endothermy. The degree of endothermy varies from the billfishes, which warm only their eyes and brain, to the bluefin tuna and the porbeagle shark, which maintain body temperatures in excess of 20 °C (68 °F) above ambient water temperatures.

Endothermy, though metabolically costly, is thought to provide advantages such as increased muscle strength, higher rates of central nervous system processing, and higher rates of digestion.

Reproductive System

Fish reproductive organs include testicles and ovaries. In most species, gonads are paired organs of similar size, which can be partially or totally fused. There may also be a range of secondary organs that increase reproductive fitness.

In terms of spermatogonia distribution, the structure of teleosts testes has two types: in the most common, spermatogonia occur all along the seminiferous tubules, while in atherinomorph fish they are confined to the distal portion of these structures. Fish can present cystic or semi-cystic spermatogenesis in relation to the release phase of germ cells in cysts to the seminiferous tubules lumen.

Fish ovaries may be of three types: gymnovarian, secondary gymnovarian or cystovarian. In the first type, the oocytes are released directly into the coelomic cavity and then enter the ostium, then through the oviduct and are eliminated. Secondary gymnovarian ovaries shed ova into the coelom from which they go directly into the oviduct. In the third type, the oocytes are conveyed to the exterior through the oviduct. Gymnovaries are the primitive condition found in lungfish, sturgeon, and bowfin. Cystovaries characterize most teleosts, where the ovary lumen has continuity with the oviduct. Secondary gymnovaries are found in salmonids and a few other teleosts.

Oogonia development in teleosts fish varies according to the group, and the determination of oogenesis dynamics allows the understanding of maturation and fertilization processes. Changes in the nucleus, ooplasm, and the surrounding layers characterize the oocyte maturation process.

Postovulatory follicles are structures formed after oocyte release; they do not have endocrine function, present a wide irregular lumen, and are rapidly reabsorbed in a process involving the apoptosis of follicular cells. A degenerative process called follicular atresia reabsorbs vitellogenic oocytes not spawned. This process can also occur, but less frequently, in oocytes in other development stages.

Some fish, like the California sheephead, are hermaphrodites, having both testes and ovaries either at different phases in their life cycle or, as in hamlets, have them simultaneously.

Over 97% of all known fish are oviparous, that is, the eggs develop outside the mother's body. Examples of oviparous fish include salmon, goldfish, cichlids, tuna, and eels. In the majority of these species, fertilisation takes place outside the mother's body, with the male and female fish shedding their gametes into the surrounding water. However, a few oviparous fish practice internal fertilization, with the male using some sort of intromittent organ to deliver sperm into the genital opening of the female, most notably the oviparous sharks, such as the horn shark, and oviparous rays, such as skates. In these cases, the male is equipped with a pair of modified pelvic fins known as claspers.

Marine fish can produce high numbers of eggs which are often released into the open water column. The eggs have an average diameter of 1 millimetre (0.039 in).

Organs: 1. Liver, 2. Gas bladder, 3. Roe, 4. Pyloric caeca, 5. Stomach, 6. Intestine.

Egg of lamprey.

Egg of catshark (mermaids' purse).

Egg of bullhead shark.

Egg of chimaera.

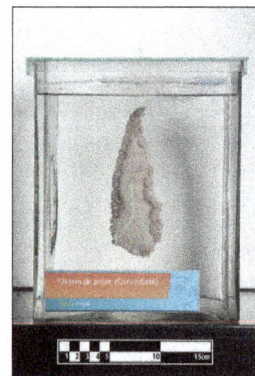

Ovary of fish (Corumbatá).

The newly hatched young of oviparous fish are called larvae. They are usually poorly formed, carry a large yolk sac (for nourishment), and are very different in appearance from juvenile and adult specimens. The larval period in oviparous fish is relatively short (usually only several weeks), and larvae rapidly grow and change appearance and structure (a process termed metamorphosis) to become juveniles. During this transition larvae must switch from their yolk sac to feeding on zooplankton prey, a process which depends on typically inadequate zooplankton density, starving many larvae.

In ovoviviparous fish the eggs develop inside the mother's body after internal fertilization but receive little or no nourishment directly from the mother, depending instead on the yolk. Each embryo develops in its own egg. Familiar examples of ovoviviparous fish include guppies, angel sharks, and coelacanths.

Soe species of fish are viviparous. In such species the mother retains the eggs and nourishes the embryos. Typically, viviparous fish have a structure analogous to the placenta seen in mammals connecting the mother's blood supply with that of the embryo. Examples of viviparous fish include the surf-perches, splitfins, and lemon shark. Some viviparous fish exhibit oophagy, in which the developing embryos eat other eggs produced by the mother. This has been observed primarily among sharks, such as the shortfin mako and porbeagle, but is known for a few bony fish as well, such as the halfbeak Nomorhamphus ebrardtii. Intrauterine cannibalism is an even more unusual mode of vivipary, in which the largest embryos eat weaker and smaller siblings. This behavior is also most commonly found among sharks, such as the grey nurse shark, but has also been reported for Nomorhamphus ebrardtii.

Aquarists commonly refer to ovoviviparous and viviparous fish as livebearers.

BIRD

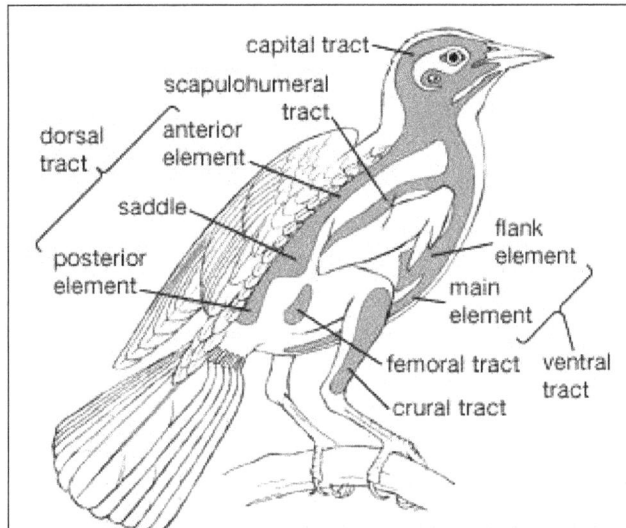

Basic body feather tracts on a generalized songbird. The shaded a
reas show the right half of each tract.

Birds (class Aves), are the unique living species having feathers, the major characteristic that distinguishes them from all other animals. A more-elaborate definition would note that they are warm-blooded vertebrates more related to reptiles than to mammals and that they have a four-chambered heart (as do mammals), forelimbs modified into wings (a trait shared with bats), a hard-shelled egg, and keen vision, the major sense they rely on for information about the environment. Their sense of smell is not highly developed, and auditory range is limited. Most birds are diurnal in habit. More than 1,000 extinct species have been identified from fossil remains.

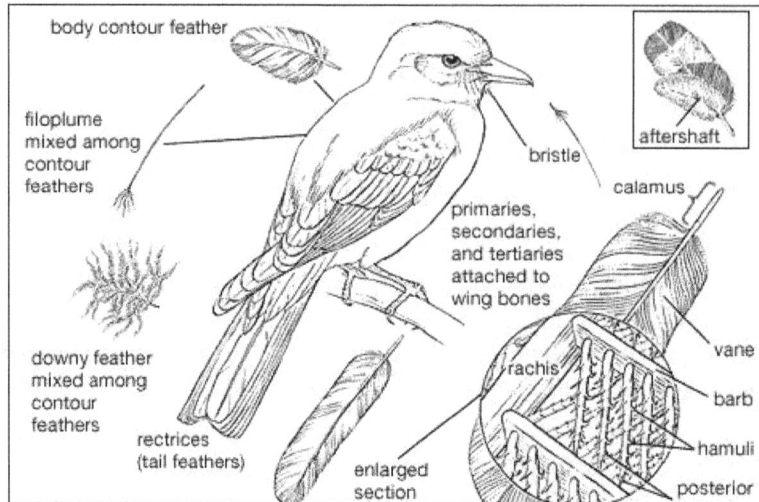

Feather types and their distribution on a typical perching bird.

Since earliest times birds have been not only a material but also a cultural resource. Bird figures were created by prehistoric humans in the Lascaux Grotto of France and have featured prominently in the mythology and literature of societies throughout the world. Long before ornithology was practiced as a science, interest in birds and the knowledge of them found expression in conversation and stories, which then crystallized into the records of general culture. Ancient Egyptian hieroglyphs and paintings, for example, include bird figures. The Bible refers to Noah's use of the raven and dove to bring him information about the proverbial Flood.

Various bird attributes, real or imagined, have led to their symbolic use in language as in art. Aesop's fables abound in bird characters. The Physiologus and its descendants, the bestiaries of the Middle Ages, contain moralistic writings that use birds as symbols for conveying ideas but indicate little knowledge of the birds themselves. Supernatural beliefs about birds probably took hold as early as recognition of the fact that some birds were good to eat. Australian Aborigines, for example, drove the black-and-white flycatcher from camp, lest it overhear conversation and carry the tales to enemies. Peoples of the Pacific Islands saw frigate birds as symbols of the Sun and as carriers of omens and frequently portrayed them in their art. The raven—a common symbol of dark prophecy—was the most important creature to the Indians of the Pacific Northwest and was immortalized in Edgar Allan Poe's poem "The Raven." Eagles have long been symbols of power and prestige in many parts of the world, including Europe, where their representations are often seen in heraldry. Native Americans sprinkled eagle down before guests as a sign of peace and friendship, and eagle feathers were commonly used in rituals and headdresses. The resplendent quetzal—the national bird of Guatemala, which shares its name with the currency and is a popular motif in art, fabric, and jewelry—was worshipped and deified by the ancient Mayans and Aztecs. Highly symbolic birds include the phoenix, representing resurrection, and the owl, a common symbol of wisdom but also a reminder of death in Native American mythology. The bird in general has long been a common Christian symbol of the transcendent soul, and in medieval iconography a bird entangled in foliage symbolized the soul embroiled in the materialism of the secular world.

In modern times the recreational pleasures of bird-watching have grown in tandem with the rise of environmentalism. Evolving from the American and European "shoot-and-stuff" mania of the 19th century, bird-watching became a sportlike activity based on rapid identification—the rarest being the most rewarding—with the aid of binoculars and spotting scopes. The change from shooting to sighting coincided with campaigns, beginning about 1900, to halt the slaughter of wild birds for food and millinery. Bird-watching was advanced by the publication of excellent field guides and improvements in photography and sound recording. By mid-century the watcher's enjoyable but rather unsophisticated tallying of "year lists" and "life lists" of species personally observed was being augmented, if not replaced, by interest in careful studies of bird Behavior, migration, ecology, and the like. This trend was abetted by bird banding (called ringing in the United Kingdom) and by such organizations as the British Trust for Ornithology and the National Audubon Society, which coordinate professional and amateur observations and efforts with scientific studies.

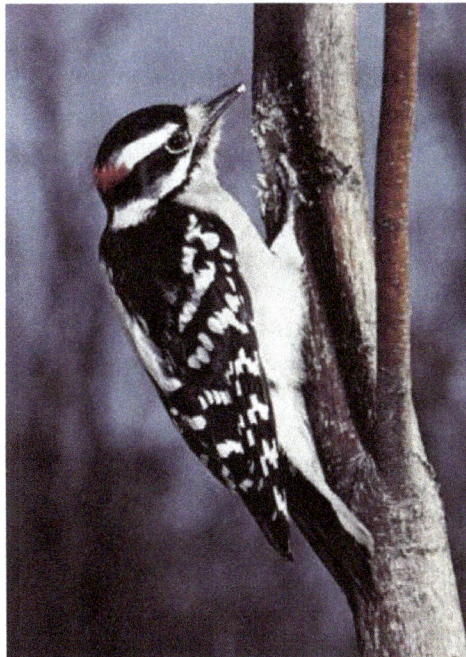

Downy woodpecker (Dendrocopos pubescens).

Features

Birds arose as warm-blooded, arboreal, flying creatures with forelimbs adapted for flight and hind limbs for perching. This basic plan has become so modified during the course of evolution that in some forms it is difficult to recognize.

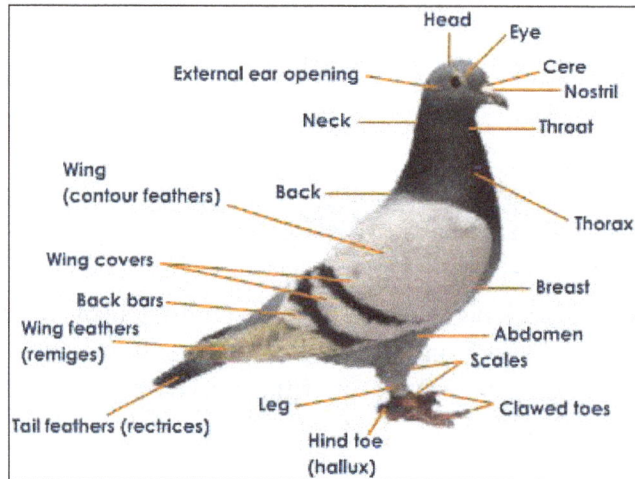

Pigeon.

Among flying birds, the wandering albatross has the greatest wingspan, up to 3.5 metres (11.5 feet), and the trumpeter swan perhaps the greatest weight, 17 kg (37 pounds). In the largest flying birds, part of the bone is replaced by air cavities (pneumatic skeletons) because the maximum size attainable by flying birds is limited by the fact that wing area varies as the square of linear proportions, and weight or volume as the cube. During the Pleistocene Epoch (2.6 million to 11,700 years ago) lived a bird called Teratornis incredibilis. Though similar to the condors of today, it had a larger estimated wingspan of about 5 metres (16.5 feet) and was by far the largest known flying bird.

The smallest living bird is generally acknowledged to be the bee hummingbird of Cuba, which is 6.3 cm (2.5 inches) long and weighs less than 3 grams (about 0.1 ounce). The minimum size is probably governed by another aspect of the surface-volume ratio: the relative increase, with decreasing size, in surface through which heat can be lost. The small size of some hummingbirds may be facilitated by a decrease in heat loss resulting from their becoming torpid at night.

When birds lose the power of flight, the limit on their maximum size is increased, as can be seen in the ostrich and other ratites such as the emu, cassowary, and rhea. The ostrich is the largest living bird and may stand 2.75 metres (9 feet) tall and weigh 150 kg (330 pounds). Some recently extinct birds were even larger: the largest moas of New Zealand and the elephant birds of Madagascar may have reached over 3 metres (10 feet) in height.

The ability to fly has permitted an almost unlimited diversification of birds, so that they are now found virtually everywhere on Earth, from occasional stragglers over the polar ice caps to complex communities in tropical forests. In general the number of species found breeding in a given area is directly proportional to the size of the area and the diversity of habitats available. The total number of species is also related to such factors as the position of the area with respect to migration routes and to wintering grounds of species that nest outside the area. In the United States, Texas and California have the most—approximately 620 for each (the figure varies based on criteria used for inclusion on state lists, such as unconfirmed, accidental, hypothetical, extirpated, and extinct species). More than 920 species have been recorded from North America north of Mexico. The figure for Europe west of the Ural Mountains and including most of Turkey is 514. More than 700 species live in Russia. At least 4,400 species live in North and South America. Although several South American countries boast well over 1,000 species, Costa Rica, with an area of only about 51,000 square km (about 20,000 square miles) and a known avifauna of more than 800 species, probably has the most diversity for its size. Asia accounts for more than 25 percent of the world's species, with 2,700 species, and Africa slightly less, with about 2,300.

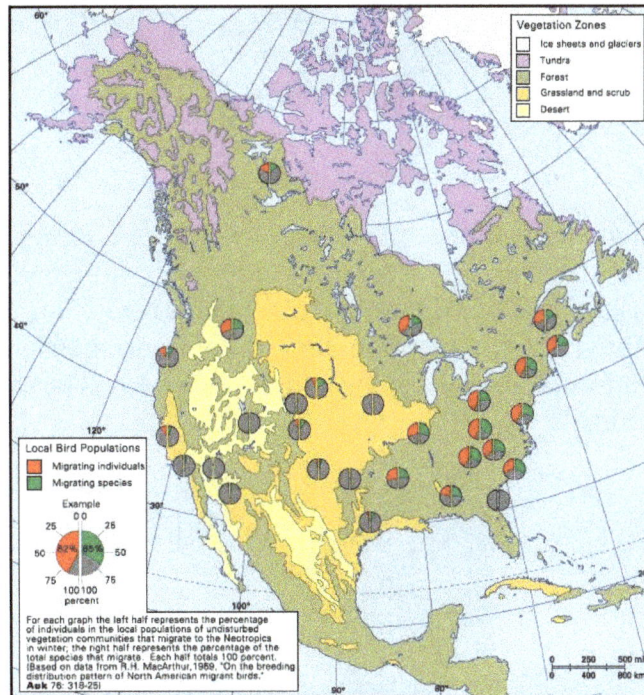

Proportion of North American birds that migrate to the tropics.

In addition to their importance in literature and legend, birds have been significant to human society in myriad ways. Birds and their eggs have been at least incidental sources of food for humans since their origin and still are in most societies. The eggs of some colonial seabirds, such as gulls, terns, and murres, or guillemots, and the young of some muttonbirds are even now harvested in large quantities. With the development of agrarian human cultures, several species of chickens, ducks, geese, and pigeons were taken in early and have been selectively bred into many varieties. These domestic birds are descended, respectively, from the red jungle fowl (Gallus gallus), mallard duck (Anas platyrhynchos), greylag goose (Anser anser), and rock dove (Columba livia). After the discovery of the New World, the turkey (Meleagris gallopavo), which had already been domesticated by the Indians, and the Muscovy duck (Cairina moschata) were brought to Europe and produced several varieties. Guinea fowl (Numida meleagris) from Africa were also widely exported and kept not only for food but also because they are noisy when alarmed, thus warning of the approach of intruders.

Besides being a food source, pigeons have long been bred and trained for carrying messages, their wartime use dating to the Roman era, according to Pliny the Elder. Messenger pigeons were widely used by German, British, and American forces in World Wars I and II and by the United States in the Korean War. In the South Seas, the ability of frigate birds to "home" to their nesting colonies enabled island inhabitants to send messages by these birds.

Red jungle fowl (Gallus gallus).

With the development of modern culture, hunting evolved from a foraging activity to a sport, in which the food value of the game became secondary. Large sums are now spent annually on hunting waterfowl, quail, grouse, pheasants, doves, and other game birds. Sets of rules and conventions have been set up for hunting, and in one elaborate form of hunting, falconry, there is not only a large body of specialized information on keeping and training falcons but also a complex terminology, much of it centuries old.

California quail (Callipepla californica).

Feathers have been used for decoration for many thousands of years. Their use in the headpieces of indigenous peoples throughout the world is well known. Feather robes were made by Polynesians and Eskimos; and down quilts, mattresses, and pillows are part of traditional European folk culture. Large feathers have often been used in fans, thereby providing an example of an object put to opposite uses—for cooling as well as for conserving heat. Whereas most feathers used in decorating are now saved as by-products of poultry raising or hunting, until early in the 20th century, egrets, grebes, and other birds were widely shot for their plumes alone. Ostrich farms have been established to produce plumes as well as meat, and some ostriches have been raised specifically for racing. Large quills were once widely used for writing, and feathers have long been used on arrows and fishing lures.

Hawaiian royal cloak made of netting
into which feathers are knotted.

Many birds are kept as pets. Small finches and parrots are especially popular and easy to keep. Of these, the canary (Serinus canaria) and the budgerigar of Australia (Melopsittacus undulatus, often called a parakeet) are widely kept and have been bred for a variety of colour types. On large parks and estates, ornamental species such as peacocks (Pavo cristatus), swans, and various exotic waterfowl and pheasants are often kept. Zoological parks in many cities import birds from many lands and are a source of recreation and enjoyment for millions of people each year.

The canary (Serinus canaria), a member of the finch family, is
one of the most popular pet songbirds in the world.

With the rise of agriculture, man's relationship with birds became more complex. Vast quantities of guano (bird excrement) were mined from island breeding colonies for use as fertilizer for crops. However, in regions where grain and fruit are grown, depredations by birds may be a serious problem. In North America various species of blackbirds (family Icteridae) are serious pests in grainfields; in Africa a grain-eating finch, the red-billed quelea (Quelea quelea), occurs like locusts, in plague proportions so numerous that alighting flocks may break the branches of trees. The use of city buildings for roosts by large flocks of starlings and blackbirds is also a problem, as is the nesting of albatrosses on airplane runways on Pacific islands. As a result of these problems, conferences on the control of avian pests are commonly held.

Common starling (Sturnus vulgaris).

Although birds are subject to a great range of diseases and parasites, only a few of these are known to be capable of infecting man. Notable exceptions are ornithosis psittacosis, or parrot fever, a serious and sometimes fatal disease resembling viral pneumonia. The microorganism responsible for the disease is transmitted directly to man from pigeons, parrots, and a variety of other birds via their excrement. Encephalitis, an inflammation of the brain, is also serious, but this infection is transmitted to man and to his domestic animals via biting arthropods, including mosquitoes. West Nile virus can likewise be transmitted. Wild birds may also act as reservoirs for diseases that adversely affect domesticated birds.

Flock of red-billed queleas (Quelea quelea),
Etosha National Park, Namibia.

The study of birds has contributed much to both the theoretical and practical aspects of biology. Charles Darwin's studies of the Galapagos finches and other birds during the voyage of HMS Beagle were important in his formulation of the idea of the origin of species through natural selection. Collections of birds in research museums still provide the bases for important studies of geographic variation, speciation, and zoogeography, because birds are one of the best known of animal groups. Early work on the domestic fowl added to the development of both genetics and embryology. The study of animal Behavior (ethology) has been based to a large extent on studies of birds by Konrad Lorenz, Nikolaas Tinbergen, and their successors. Birds also have been the primary group in the study of migration and orientation and the effect of hormones on Behavior and physiology.

Man's impact on bird populations is very strong. Since 1680 approximately 80 species of birds have become extinct, and a larger number are seriously endangered. While pollution and pesticides are important factors in the decline of certain large species, such as the peregrine falcon, osprey, and California condor, the destruction of natural areas and introduction of exotic animals and diseases have probably been the most devastating. Concerted efforts of research and conservation are required to ensure the survival of rare species.

The dusky seaside sparrow, which went extinct.

Classification

Distinguishing Taxonomic Features

In classifying birds, most systematists have historically relied upon structural characteristics to infer evolutionary relationships. Plumage characteristics include the number of various feather types; the presence or absence of down on the feather tracts and on the preen gland; and the presence or absence of an aftershaft. Characteristics of the bill and feet are also useful, as is the arrangement of bones in the palate and around the nostrils. The presence or absence of certain thigh muscles is considered, as are the arrangement of the carotid arteries, the syrinx, and the deep flexor tendons of the toes as well as the condition of the young when hatched. Advances in the study of DNA sequences and computerized construction of phylogenetic trees have provided new means of testing hypotheses of taxonomic relationships.

Critical Appraisal

It has frequently been stated that birds are one of the best known of animal groups. This is true in the sense that most of the living species and subspecies in the world have probably been described; but because of inadequacies in the fossil record and repeated cases of convergent evolution within the group, our knowledge of the phylogenetic relationships between orders, suborders, and families of birds is inferior to that of mammals and reptiles. Most, if not all, of the major lineages of modern birds arose rapidly in the Late Cretaceous and the Paleogene Period (about 100 million to 23 million years ago). DNA data continue to resolve the relationships among major groups of birds. The penguins (Sphenisciformes), tube-nosed seabirds (Procellariiformes), and pelicans (Pelecaniformes) form a triad of related lineages. Waterfowl (Anseriformes) and chickenlike birds (Galliformes) are linked and together may be the oldest assemblage of modern birds. Some caprimulgiforms (owlet frogmouths) seem clearly related to swifts (Apodiformes) through a link between owlet frogmouths and treeswifts.

The taxonomic positions of several bird groups remain open to question. The hoatzin, included below in the Cuculiformes, is often given its own order, Opisthocomiformes. The sandgrouse are listed separately in order Pteroclidiformes. The turacos, sometimes included in the Cuculiformes, are considered by many authors to warrant separation and are listed here as Musophagiformes. Diatryma and several related genera of extinct flightless predators are often placed in a distinct order, Diatrymiformes, near Gruiformes. The flamingos, which constitute the order Phoenicopteriformes in some classifications, are placed in the Ciconiiformes in this classification, but their relationships are still unknown.

One area particularly in need of study is the relationships among the various groups of ratites (ostriches, rheas, emus, moas, and others). Formerly, some authorities argued that these birds and the penguins arose independently from cursorial reptiles, but it is now generally agreed that all of them passed through a flying stage in the course of their evolution. The ratite groups differ greatly in morphology and yet show remarkable similarities in palate and bill characters. The principal unanswered questions are how many different flightless lines evolved from flying ancestors and from how many different groups the flying ancestors evolved. On zoogeographic grounds, it is likely that the isolated kiwi-moa, elephant bird, and emu-cassowary lines arose independently from each other and from ratites on the other continents. But the ostriches and rheas could be descended from a common flightless ancestor because of the known former land connections from Asia to North and South America. Kiwis, ostriches, rheas, emus, and cassowaries are contained within order Struthioniformes in this classification.

The evolutionary sequence of the bird orders starts with ratites and marine seabirds and ends with songbirds. Beginning in the 1980s, Charles Sibley proposed radically different listings of the nonpasserine orders on the basis of his pioneering DNA analyses.

REPTILES

Reptiles are tetrapod animals in the class Reptilia, comprising today's turtles, crocodilians, snakes, amphisbaenians, lizards, tuatara, and their extinct relatives. The study of these traditional reptile orders, historically combined with that of modern amphibians, is called herpetology.

Because some reptiles are more closely related to birds than they are to other reptiles (e.g., crocodiles are more closely related to birds than they are to lizards), the traditional groups of "reptiles" listed above do not together constitute a monophyletic grouping or clade (consisting of all descendants of a common ancestor). For this reason, many modern scientists prefer to consider the birds part of Reptilia as well, thereby making Reptilia a monophyletic class, including all living Diapsids.

The earliest known proto-reptiles originated around 312 million years ago during the Carboniferous period, having evolved from advanced reptiliomorph tetrapods that became increasingly adapted to life on dry land. Some early examples include the lizard-like Hylonomus and Casineria. In addition to the living reptiles, there are many diverse groups that are now extinct, in some cases due to mass extinction events. In particular, the Cretaceous–Paleogene extinction event wiped out the pterosaurs, plesiosaurs, ornithischians, and sauropods, as well as many species of theropods, including troodontids, dromaeosaurids, tyrannosaurids, and abelisaurids, along with many Crocodyliformes, and squamates (e.g. mosasaurids).

Modern non-avian reptiles inhabit all the continents except Antarctica, although some birds are found on the periphery of Antarctica. Several living subgroups are recognized: Testudines (turtles and tortoises), 350 species; Rhynchocephalia (tuatara from New Zealand), 1 species; Squamata (lizards, snakes, and worm lizards), over 10,200 species; Crocodilia (crocodiles, gavials, caimans, and alligators), 24 species; and Aves (birds), approximately 10,000 species.

Reptiles are tetrapod vertebrates, creatures that either have four limbs or, like snakes, are descended from four-limbed ancestors. Unlike amphibians, reptiles do not have an aquatic larval stage. Most reptiles are oviparous, although several species of squamates are viviparous, as were some extinct aquatic clades—the fetus develops within the mother, contained in a placenta rather than an eggshell. As amniotes, reptile eggs are surrounded by membranes for protection and transport, which adapt them to reproduction on dry land. Many of the viviparous species feed their fetuses through various forms of placenta analogous to those of mammals, with some providing initial care for their hatchlings. Extant reptiles range in size from a tiny gecko, Sphaerodactylus ariasae, which can grow up to 17 mm (0.7 in) to the saltwater crocodile, Crocodylus porosus, which can reach 6 m (19.7 ft) in length and weigh over 1,000 kg (2,200 lb).

Classification

Phylogenetics and Modern Definition

By the early 21st century, vertebrate paleontologists were beginning to adopt phylogenetic taxonomy, in which all groups are defined in such a way as to be monophyletic; that is, groups include all descendants of a particular ancestor. The reptiles as historically defined are paraphyletic, since they exclude both birds and mammals. These respectively evolved from dinosaurs and from early therapsids, which were both traditionally called reptiles. Birds are more closely related to crocodilians than the latter are to the rest of extant reptiles. Colin Tudge wrote:

> "Mammals are a clade, and therefore the cladists are happy to acknowledge the traditional taxon Mammalia; and birds, too, are a clade, universally ascribed to the formal taxon Aves. Mammalia and Aves are, in fact, subclades within the grand clade of the Amniota. But the traditional class Reptilia is not a clade. It is just a section of the clade Amniota: the section that is left after the Mammalia and Aves have been hived off. It cannot be defined by synapomorphies, as is the proper way. Instead, it is defined by a combination of the features it has and the features it lacks: reptiles are the amniotes that lack fur or feathers. At best, the cladists suggest, we could say that the traditional Reptilia are 'non-avian, non-mammalian amniotes'."

Despite the early proposals for replacing the paraphyletic Reptilia with a monophyletic Sauropsida, which includes birds, that term was never adopted widely or, when it was, was not applied consistently. When Sauropsida was used, it often had the same content or even the same definition as Reptilia. In 1988, Jacques Gauthier proposed a cladistic definition of Reptilia as a monophyletic node-based crown group containing turtles, lizards and snakes, crocodilians, and birds, their common ancestor and all its descendants. Because the actual relationship of turtles to other reptiles was not yet well understood at this time, Gauthier's definition came to be considered inadequate.

A variety of other definitions were proposed by other scientists in the years following Gauthier's paper. The first such new definition, which attempted to adhere to the standards of the

PhyloCode, was published by Modesto and Anderson in 2004. Modesto and Anderson reviewed the many previous definitions and proposed a modified definition, which they intended to retain most traditional content of the group while keeping it stable and monophyletic. They defined Reptilia as all amniotes closer to Lacerta agilis and Crocodylus niloticus than to Homo sapiens. This stem-based definition is equivalent to the more common definition of Sauropsida, which Modesto and Anderson synonymized with Reptilia, since the latter is better known and more frequently used. Unlike most previous definitions of Reptilia, however, Modesto and Anderson's definition includes birds, as they are within the clade that includes both lizards and crocodiles.

Phylogenetic classifications group the traditional "mammal-like reptiles", like this Varanodon, with other synapsids, not with extant reptiles.

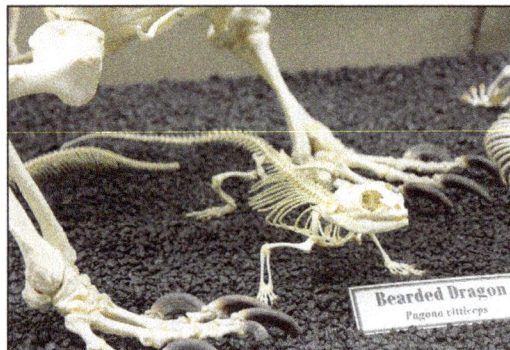

Bearded dragon (pogona) skeleton on display at the Museum of Osteology.

Taxonomy

Classification to order level of the reptiles, after Benton, 2014.

- Class Reptilia,
- Subclass Parareptilia,
- Order Pareiasauromorpha,
- Subclass Eureptiliam,
- Infraclass Diapsida,
- Order Younginiformes,
- Infraclass Neodiapsida,
- Order Testudinata (turtles),

- Infraclass Lepidosauromorpha,

- Infrasubclass Unnamed,

- Infraclass Ichthyosauria,

- Order Thalattosauria,

- Superorder Lepidosauriformes,

- Order Rhynchocephalia (tuatara),

- Order Squamata (lizards & snakes),

- Infrasubclass Sauropterygia,

- Order Placodontia,

- Order Eosauropterygia,

- Order Plesiosauria,

- Infraclass Archosauromorpha,

- Order Rhynchosauria,

- Order Protorosauria,

- Order Phytosauria,

- Division Archosauriformes,

- Subdivision Archosauria,

- Superorder Crocodylomorpha,

- Order Crocodilia,

- Infradivision Avemetatarsalia,

- Infrasubdivision Ornithodira,

- Order Pterosauria,

- Superorder Dinosauria,

- Order Saurischia (incl. Clade Aves),

- Order Ornithischia.

Phylogeny

The cladogram presented here illustrates the "family tree" of reptiles, and follows a simplified version of the relationships found by M.S. Lee, in 2013. All genetic studies have supported the hypothesis that turtles are diapsids; some have placed turtles within archosauriformes, though a few have recovered turtles as lepidosauriformes instead. The cladogram below used a combination of genetic (molecular) and fossil (morphological) data to obtain its results.

The Position of Turtles

The placement of turtles has historically been highly variable. Classically, turtles were considered to be related to the primitive anapsid reptiles. Molecular work has usually placed turtles within the diapsids. As of 2013, three turtle genomes have been sequenced. The results place turtles as a sister clade to the archosaurs, the group that includes crocodiles, dinosaurs, and birds.

Morphology and Physiology

Circulation

All squamates and turtles have a three-chambered heart consisting of two atria, one variably partitioned ventricle, and two aortas that lead to the systemic circulation. The degree of mixing of oxygenated and deoxygenated blood in the three-chambered heart varies depending on the species and physiological state. Under different conditions, deoxygenated blood can be shunted back to the body or oxygenated blood can be shunted back to the lungs. This variation in blood flow has been hypothesized to allow more effective thermoregulation and longer diving times for aquatic species, but has not been shown to be a fitness advantage.

For example, Iguana hearts, like the majority of the squamates hearts, are composed of three chambers with two aorta and one ventricle, cardiac involuntary muscles. The main structures of the heart are the sinus venosus, the pacemaker, the left atrium, the right atruim, the atrioventricular valve, the cavum venosum, cavum arteriosum, the cavum pulmonale, the muscular ridge, the ventricular ridge, pulmonary veins, and paired aortic arches.

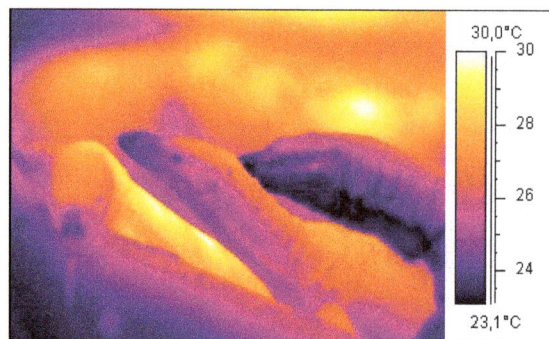

Thermographic image of monitor lizards.

Some squamate species (e.g., pythons and monitor lizards) have three-chambered hearts that become functionally four-chambered hearts during contraction. This is made possible by a muscular ridge that subdivides the ventricle during ventricular diastole and completely divides it during ventricular systole. Because of this ridge, some of these squamates are capable of producing ventricular pressure differentials that are equivalent to those seen in mammalian and avian hearts.

Crocodilians have an anatomically four-chambered heart, similar to birds, but also have two systemic aortas and are therefore capable of bypassing their pulmonary circulation.

Metabolism

Modern non-avian reptiles exhibit some form of cold-bloodedness (i.e. some mix of poikilothermy, ectothermy, and bradymetabolism) so that they have limited physiological means of keeping the body temperature constant and often rely on external sources of heat. Due to a less stable core temperature than birds and mammals, reptilian biochemistry requires enzymes capable of maintaining efficiency over a greater range of temperatures than in the case for warm-blooded animals. The optimum body temperature range varies with species, but is typically below that of warm-blooded animals; for many lizards, it falls in the 24°–35 °C (75°–95 °F) range, while extreme heat-adapted species, like the American desert iguana Dipsosaurus dorsalis, can have optimal physiological temperatures in the mammalian range, between 35° and 40 °C (95° and 104 °F). While the optimum temperature is often encountered when the animal is active, the low basal metabolism makes body temperature drop rapidly when the animal is inactive.

As in all animals, reptilian muscle action produces heat. In large reptiles, like leatherback turtles, the low surface-to-volume ratio allows this metabolically produced heat to keep the animals warmer than their environment even though they do not have a warm-blooded metabolism. This form of homeothermy is called gigantothermy; it has been suggested as having been common in large dinosaurs and other extinct large-bodied reptiles.

The benefit of a low resting metabolism is that it requires far less fuel to sustain bodily functions. By using temperature variations in their surroundings, or by remaining cold when they do not need to move, reptiles can save considerable amounts of energy compared to endothermic animals of the same size. A crocodile needs from a tenth to a fifth of the food necessary for a lion of the same weight and can live half a year without eating. Lower food requirements and adaptive metabolisms allow reptiles to dominate the animal life in regions where net calorie availability is too low to sustain large-bodied mammals and birds.

It is generally assumed that reptiles are unable to produce the sustained high energy output necessary for long distance chases or flying. Higher energetic capacity might have been responsible for the evolution of warm-bloodedness in birds and mammals. However, investigation of correlations between active capacity and thermophysiology show a weak relationship. Most extant reptiles are carnivores with a sit-and-wait feeding strategy; whether reptiles are cold blooded due to their ecology is not clear. Energetic studies on some reptiles have shown active capacities equal to or greater than similar sized warm-blooded animals.

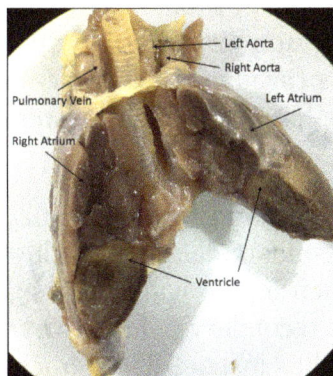

A juvenile Iguana heart bisected through the ventricle, bisecting the left and right atrium.

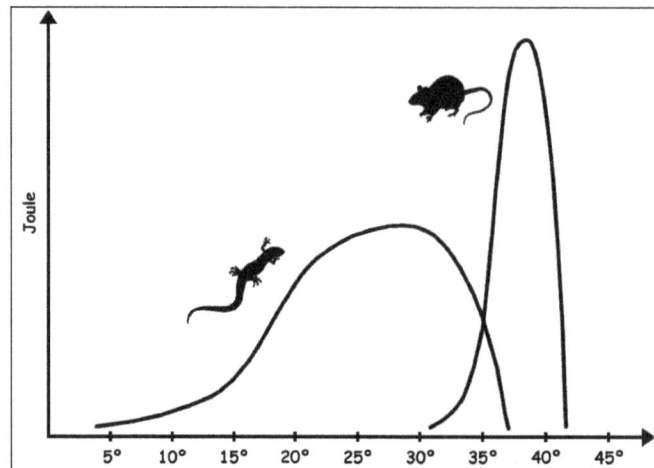

Sustained energy output (joules) of a typical reptile versus a similar size mammal as a function of core body temperature. The mammal has a much higher peak output, but can only function over a very narrow range of body temperature.

Respiratory System

All reptiles breathe using lungs. Aquatic turtles have developed more permeable skin, and some species have modified their cloaca to increase the area for gas exchange. Even with these adaptations, breathing is never fully accomplished without lungs. Lung ventilation is accomplished differently in each main reptile group. In squamates, the lungs are ventilated almost exclusively by the axial musculature. This is also the same musculature that is used during locomotion. Because of this constraint, most squamates are forced to hold their breath during intense runs. Some, however, have found a way around it. Varanids, and a few other lizard species, employ buccal pumping as a complement to their normal "axial breathing". This allows the animals to completely fill their lungs during intense locomotion, and thus remain aerobically active for a long time. Tegu lizards are known to possess a proto-diaphragm, which separates the pulmonary cavity from the visceral cavity. While not actually capable of movement, it does allow for greater lung inflation, by taking the weight of the viscera off the lungs.

Crocodilians actually have a muscular diaphragm that is analogous to the mammalian diaphragm. The difference is that the muscles for the crocodilian diaphragm pull the pubis (part of the pelvis, which is movable in crocodilians) back, which brings the liver down, thus freeing space for the lungs to expand. This type of diaphragmatic setup has been referred to as the "hepatic piston". The airways form a number of double tubular chambers within each lung. On inhalation and exhalation air moves through the airways in the same direction, thus creating a unidirectional airflow through the lungs. A similar system is found in birds, monitor lizards and iguanas.

Most reptiles lack a secondary palate, meaning that they must hold their breath while swallowing. Crocodilians have evolved a bony secondary palate that allows them to continue breathing while remaining submerged (and protect their brains against damage by struggling prey). Skinks (family Scincidae) also have evolved a bony secondary palate, to varying degrees. Snakes took a different approach and extended their trachea instead. Their tracheal extension sticks out like a fleshy straw, and allows these animals to swallow large prey without suffering from asphyxiation.

Turtles and Tortoises

How turtles and tortoises breathe has been the subject of much study. To date, only a few species have been studied thoroughly enough to get an idea of how those turtles breathe. The varied results indicate that turtles and tortoises have found a variety of solutions to this problem.

The difficulty is that most turtle shells are rigid and do not allow for the type of expansion and contraction that other amniotes use to ventilate their lungs. Some turtles, such as the Indian flapshell (Lissemys punctata), have a sheet of muscle that envelops the lungs. When it contracts, the turtle can exhale. When at rest, the turtle can retract the limbs into the body cavity and force air out of the lungs. When the turtle protracts its limbs, the pressure inside the lungs is reduced, and the turtle can suck air in. Turtle lungs are attached to the inside of the top of the shell (carapace), with the bottom of the lungs attached (via connective tissue) to the rest of the viscera. By using a series of special muscles (roughly equivalent to a diaphragm), turtles are capable of pushing their viscera up and down, resulting in effective respiration, since many of these muscles have attachment points in conjunction with their forelimbs (indeed, many of the muscles expand into the limb pockets during contraction).

Breathing during locomotion has been studied in three species, and they show different patterns. Adult female green sea turtles do not breathe as they crutch along their nesting beaches. They hold their breath during terrestrial locomotion and breathe in bouts as they rest. North American box turtles breathe continuously during locomotion, and the ventilation cycle is not coordinated with the limb movements. This is because they use their abdominal muscles to breathe during locomotion. The last species to have been studied is the red-eared slider, which also breathes during locomotion, but takes smaller breaths during locomotion than during small pauses between locomotor bouts, indicating that there may be mechanical interference between the limb movements and the breathing apparatus. Box turtles have also been observed to breathe while completely sealed up inside their shells.

Red-eared slider taking a gulp of air.

Skin

Reptilian skin is covered in a horny epidermis, making it watertight and enabling reptiles to live on dry land, in contrast to amphibians. Compared to mammalian skin, that of reptiles is rather thin and lacks the thick dermal layer that produces leather in mammals. Exposed parts of reptiles are protected by scales or scutes, sometimes with a bony base, forming armor. In lepidosaurians, such as lizards and snakes, the whole skin is covered in overlapping epidermal scales. Such scales were once thought to be typical of the class Reptilia as a whole, but are now known to occur only in lepidosaurians. The scales found in turtles and crocodiles are of dermal, rather than epidermal, origin and are properly termed scutes. In turtles, the body is hidden inside a hard shell composed of fused scutes.

Lacking a thick dermis, reptilian leather is not as strong as mammalian leather. It is used in leather-wares for decorative purposes for shoes, belts and handbags, particularly crocodile skin.

Shedding. Reptiles shed their skin through a process called ecdysis which occurs continuously throughout their lifetime. In particular, younger reptiles tend to shed once every 5–6 weeks while adults shed 3–4 times a year. Younger reptiles shed more because of their rapid growth rate. Once full size, the frequency of shedding drastically decreases. The process of ecdysis involves forming a new layer of skin under the old one. Proteolytic enzymes and lymphatic fluid is secreted between the old and new layers of skin. Consequently, this lifts the old skin from the new one allowing shedding to occur. Snakes will shed from the head to the tail while lizards shed in a "patchy pattern". Dysecdysis, a common skin disease in snakes and lizards, will occur when ecdysis, or shedding, fails. There are numerous reasons why shedding fails and can be related to inadequate humidity and temperature, nutritional deficiencies, dehydration and traumatic injuries. Nutritional deficiencies decrease proteolytic enzymes while dehydration reduces lymphatic fluids to separate the skin layers. Traumatic injuries on the other hand, form scars that will not allow new scales to form and disrupt the process of ecdysis.

Skin of a sand lizard, showing squamate reptiles iconic scales.

Excretion

Excretion is performed mainly by two small kidneys. In diapsids, uric acid is the main nitrogenous waste product; turtles, like mammals, excrete mainly urea. Unlike the kidneys of mammals and birds, reptile kidneys are unable to produce liquid urine more concentrated than their body fluid. This is because they lack a specialized structure called a loop of Henle, which is present in the nephrons of birds and mammals. Because of this, many reptiles use the colon to aid in the reabsorption of water. Some are also able to take up water stored in the bladder. Excess salts are also excreted by nasal and lingual salt glands in some reptiles.

In all reptiles the urinogenital ducts and the anus both empty into an organ called a cloaca. In some reptiles, a midventral wall in the cloaca may open into a urinary bladder, but not all. It is present in all turtles and tortoises as well as most lizards, but is lacking in the monitor lizard, the legless lizards. It is absent in the snakes, alligators, and crocodiles.

Many turtles, tortoises, and lizards have proportionally very large bladders. Charles Darwin noted that the Galapagos tortoise had a bladder which could store up to 20% of its body weight. Such adaptations are the result of environments such as remote islands and deserts where water is very scarce. Other desert-dwelling reptiles have large bladders that can store a long-term reservoir of water for up to several months and aid in osmoregulation.

Turtles have two or more accessory urinary bladders, located lateral to the neck of the urinary bladder and dorsal to the pubis, occupying a significant portion of their body cavity. Their bladder is also usually bilobed with a left and right section. The right section is located under the liver, which prevents large stones from remaining in that side while the left section is more likely to have calculi.

Digestion

Most reptiles are insectivorous or carnivorous and have simple and comparatively short digestive tracts due to meat being fairly simple to break down and digest. Digestion is slower than in mammals, reflecting their lower resting metabolism and their inability to divide and masticate their food. Their poikilotherm metabolism has very low energy requirements, allowing large reptiles like crocodiles and large constrictors to live from a single large meal for months, digesting it slowly.

While modern reptiles are predominantly carnivorous, during the early history of reptiles several groups produced some herbivorous megafauna: in the Paleozoic, the pareiasaurs; and in the Mesozoic several lines of dinosaurs. Today, turtles are the only predominantly herbivorous reptile group, but several lines of agamas and iguanas have evolved to live wholly or partly on plants.

Herbivorous reptiles face the same problems of mastication as herbivorous mammals but, lacking the complex teeth of mammals, many species swallow rocks and pebbles (so called gastroliths) to aid in digestion: The rocks are washed around in the stomach, helping to grind up plant matter. Fossil gastroliths have been found associated with both ornithopods and sauropods, though whether they actually functioned as a gastric mill in the latter is disputed. Salt water crocodiles also use gastroliths as ballast, stabilizing them in the water or helping them to dive. A dual function as both stabilizing ballast and digestion aid has been suggested for gastroliths found in plesiosaurs.

A colubrid snake, Dolichophis jugularis, eating a legless lizard, Pseudopus apodus. Most reptiles are carnivorous, and many primarily eat other reptiles.

Nerves

The reptilian nervous system contains the same basic part of the amphibian brain, but the reptile cerebrum and cerebellum are slightly larger. Most typical sense organs are well developed with certain exceptions, most notably the snake's lack of external ears (middle and inner ears are present). There are twelve pairs of cranial nerves. Due to their short cochlea, reptiles use electrical tuning to expand their range of audible frequencies.

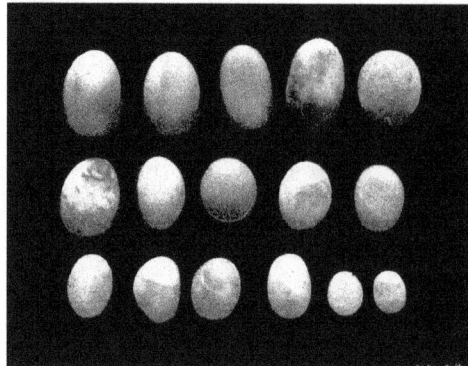

Gastroliths from a plesiosaur.

Intelligence

Reptiles are generally considered less intelligent than mammals and birds. The size of their brain relative to their body is much less than that of mammals, the encephalization quotient being about one tenth of that of mammals, though larger reptiles can show more complex brain development. Larger lizards, like the monitors, are known to exhibit complex behavior, including cooperation. Crocodiles have relatively larger brains and show a fairly complex social structure. The Komodo dragon is even known to engage in play, as are turtles, which are also considered to be social creatures, and sometimes switch between monogamy and promiscuity in their sexual behavior. One study found that wood turtles were better than white rats at learning to navigate mazes.

Vision

Most reptiles are diurnal animals. The vision is typically adapted to daylight conditions, with color vision and more advanced visual depth perception than in amphibians and most mammals. In some species, such as blind snakes, vision is reduced.

Some snakes have extra sets of visual organs (in the loosest sense of the word) in the form of pits sensitive to infrared radiation (heat). Such heat-sensitive pits are particularly well developed in the pit vipers, but are also found in boas and pythons. These pits allow the snakes to sense the body heat of birds and mammals, enabling pit vipers to hunt rodents in the dark.

Reproduction

Reptiles generally reproduce sexually, though some are capable of asexual reproduction. All reproductive activity occurs through the cloaca, the single exit/entrance at the base of the tail where waste is also eliminated. Most reptiles have copulatory organs, which are usually retracted or inverted and stored inside the body. In turtles and crocodilians, the male has a single median penis, while squamates, including snakes and lizards, possess a pair of hemipenes, only one of which is typically used in each session. Tuatara, however, lack copulatory organs, and so the male and female simply press their cloacas together as the male discharges sperm.

Most reptiles lay amniotic eggs covered with leathery or calcareous shells. An amnion, chorion, and allantois are present during embryonic life. The eggshell (1) protects the crocodile embryo (11) and keeps it from drying out, but it is flexible to allow gas exchange. The chorion (6) aids in gas exchange between the inside and outside of the egg. It allows carbon dioxide to exit the egg and oxygen gas to enter the egg. The albumin (9) further protects the embryo and serves as a reservoir for water and protein. The allantois (8) is a sac that collects the metabolic waste produced by the embryo. The amniotic sac (10) contains amniotic fluid (12) which protects and cushions the embryo. The amnion (5) aids in osmoregulation and serves as a saltwater reservoir. The yolk sac (2) surrounding the yolk (3) contains protein and fat rich nutrients that are absorbed by the embryo via vessels (4) that allow the embryo to grow and metabolize. The air space (7) provides the embryo with oxygen while it is hatching. This ensures that the embryo will not suffocate while it is hatching. There are no larval stages of development. Viviparity and ovoviviparity have evolved in many extinct clades of reptiles and in squamates. In the latter group, many species, including all boas and most vipers, utilize this mode of reproduction. The degree of viviparity varies; some species simply retain the eggs until just before hatching, others provide maternal nourishment to supplement the yolk, and yet others lack any yolk and provide all nutrients via a structure similar to the mammalian placenta. The earliest documented case of viviparity in reptiles is the Early Permian mesosaurs, although some individuals or taxa in that clade may also have been oviparous because a putative isolated egg has also been found. Several groups of Mesozoic marine reptiles also exhibited viviparity, such as mosasaurs, ichthyosaurs, and Sauropterygia, a group that include pachypleurosaurs and Plesiosauria.

Asexual reproduction has been identified in squamates in six families of lizards and one snake. In some species of squamates, a population of females is able to produce a unisexual diploid clone of the mother. This form of asexual reproduction, called parthenogenesis, occurs in several species of gecko, and is particularly widespread in the teiids (especially Aspidocelis) and lacertids (Lacerta). In captivity, Komodo dragons (Varanidae) have reproduced by parthenogenesis.

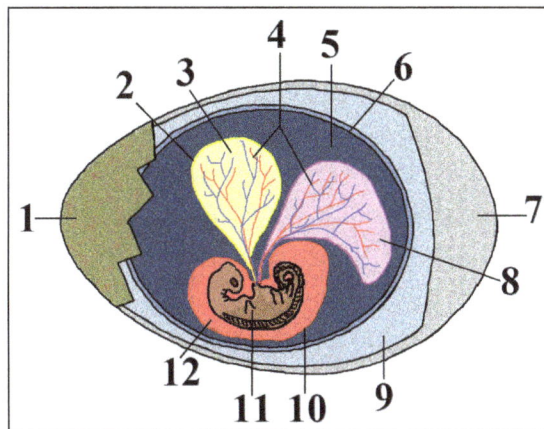

Crocodilian egg diagram 1. eggshell, 2. yolk sac, 3. yolk (nutrients), 4. vessels, 5. amnion, 6. chorion, 7. air space, 8. allantois, 9. albumin (egg white), 10. amniotic sac, 11. crocodile embryo, 12. amniotic fluid.

Parthenogenetic species are suspected to occur among chameleons, agamids, xantusiids, and typhlopids.

Some reptiles exhibit temperature-dependent sex determination (TDSD), in which the incubation temperature determines whether a particular egg hatches as male or female. TDSD is most common in turtles and crocodiles, but also occurs in lizards and tuatara. To date, there has been no confirmation of whether TDSD occurs in snakes.

Most reptiles reproduce sexually, for example this Trachylepis maculilabris skink.

Geckos mating.

Defense Mechanisms

Many small reptiles, such as snakes and lizards that live on the ground or in the water, are vulnerable to being preyed on by all kinds of carnivorous animals. Thus avoidance is the most common form of defense in reptiles. At the first sign of danger, most snakes and lizards crawl away into the undergrowth, and turtles and crocodiles will plunge into water and sink out of sight.

Camouflage and Warning

Reptiles tend to avoid confrontation through camouflage. Two major groups of reptile predators are birds and other reptiles, both of which have well developed color vision. Thus the skins of many reptiles have cryptic coloration of plain or mottled gray, green, and brown to allow them to blend into the background of their natural environment. Aided by the reptiles' capacity for remaining motionless for long periods, the camouflage of many snakes is so effective that people or domestic animals are most typically bitten because they accidentally step on them.

When camouflage fails to protect them, blue-tongued skinks will try to ward off attackers by displaying their blue tongues, and the frill-necked lizard will display its brightly colored frill. These same displays are used in territorial disputes and during courtship. If danger arises so suddenly that flight is useless, crocodiles, turtles, some lizards, and some snakes hiss loudly when confronted by an enemy. Rattlesnakes rapidly vibrate the tip of the tail, which is composed of a series of nested, hollow beads to ward of approaching danger.

In contrast to the normal drab coloration of most reptiles, the lizards of the genus Heloderma (the Gila monster and the beaded lizard) and many of the coral snakes have high-contrast warning coloration, warning potential predators they are venomous. A number of non-venomous North American snake species have colorful markings similar to those of the coral snake, an oft cited example of Batesian mimicry.

Alternative Defense in Snakes

Camouflage does not always fool a predator. When caught out, snake species adopt different defensive tactics and use a complicated set of behaviors when attacked. Some first elevate their head and spread out the skin of their neck in an effort to look large and threatening. Failure of this strategy may lead to other measures practiced particularly by cobras, vipers, and closely related species, which use venom to attack. The venom is modified saliva, delivered through fangs from a venom gland. Some non-venomous snakes, such as American hognose snakes or European grass snake, play dead when in danger; some, including the grass snake, exude a foul-smelling liquid to deter attackers.

Defense in Crocodilians

When a crocodilian is concerned about its safety, it will gape to expose the teeth and yellow tongue. If this doesn't work, the crocodilian gets a little more agitated and typically begins to make hissing sounds. After this, the crocodilian will start to change its posture dramatically to make itself look more intimidating. The body is inflated to increase apparent size. If absolutely necessary it may decide to attack an enemy.

Some species try to bite immediately. Some will use their heads as sledgehammers and literally smash an opponent, some will rush or swim toward the threat from a distance, even chasing the opponent onto land or galloping after it. The main weapon in all crocodiles is the bite, which can generate very high bite force. Many species also possess canine-like teeth. These are used primarily for seizing prey, but are also used in fighting and display.

Shedding and Regenerating Tails

Geckos, skinks, and other lizards that are captured by the tail will shed part of the tail structure through a process called autotomy and thus be able to flee. The detached tail will continue to wiggle, creating a deceptive sense of continued struggle and distracting the predator's attention from the fleeing prey animal. The detached tails of leopard geckos can wiggle for up to 20 minutes. In many species the tails are of a separate and dramatically more intense color than the rest of the body so as to encourage potential predators to strike for the tail first. In the shingleback skink and some species of geckos, the tail is short and broad and resembles the head, so that the predators may attack it rather than the more vulnerable front part.

Reptiles have amniotic eggs with hard or leathery shells, requiring internal fertilization when mating.

A camouflaged Phelsuma deubia on a palm frond.

A White-headed dwarf gecko with shed tail.

Reptiles that are capable of shedding their tails can partially regenerate them over a period of weeks. The new section will however contain cartilage rather than bone, and will never grow to the same length as the original tail. It is often also distinctly discolored compared to the rest of the body and may lack some of the external sculpting features seen in the original tail.

References

- Vertebrate, animal: britannica.com, Retrieved 20 March, 2019

- Sahney, S. & Benton, M.J. (2008). "Recovery from the most profound mass extinction of all time". Proceedings of the Royal Society B. 275 (1636): 759–765. Doi:10.1098/rspb.2007.1370. PMC 2596898. PMID 18198148

- What-are-amphibians: worldatlas.com, Retrieved 21 April, 2019

- Nelson, Joseph S. (2006). Fishes of the World (PDF) (4th ed.). John Wiley & Sons. ISBN 978-0-471-75644-6. Archived from the original (PDF) on 5 March 2013. Retrieved 30 April 2013

- Mammal, animal: britannica.com, Retrieved 22 May, 2019

- Davis, Jon R.; denardo, Dale F. (2007-04-15). "The urinary bladder as a physiological reservoir that moderates dehydration in a large desert lizard, the Gila monster Heloderma suspectum". Journal of Experimental Biology. 210 (8): 1472–1480. Doi:10.1242/jeb.003061. ISSN 0022-0949. PMID 17401130

- Bird-animal, animal: britannica.com, Retrieved 23 June, 2019

- Moyle, Peter B.; Cech, Joseph J. (2003). Fishes, An Introduction to Ichthyology (5th ed.). Benjamin Cummings. ISBN 978-0-13-100847-2

Invertebrate Zoology

The sub-discipline of zoology which studies the animals without a backbone is known as invertebrate zoology. Some of the phyla that are studied within invertebrate zoology are arthropoda, mollusca, annelida and nematoda. All the diverse aspects related to these phyla within invertebrate zoology have been carefully analyzed in this chapter.

INVERTEBRATE

Invertebrates are animals that don't have a backbone. The vertebral column is another name for the backbone. Over 90% of all species on Earth are invertebrates, and invertebrate species have been found in the fossil record as far back as 600 million years ago. Molecular biology studies suggest that all invertebrates evolved from a single invertebrate group.

Characteristics of Invertebrates

In addition to not having a backbone, invertebrates have soft bodies because they don't have an internal skeleton (endoskeleton) for support. Instead, many have structures on the outside (exoskeleton) that provide support and protection. In addition, invertebrates are cold-blooded, meaning they can't regulate their body temperature, so it changes depending on the environment.

Invertebrates are incredibly diverse. They live in fresh water, salt water, on land and as parasites in other animals. There are invertebrates that are carnivorous (meat eaters), herbivores (plant eaters) and omnivores (meat and plant eaters). There are even some invertebrate species that grow bacteria and cells inside their bodies that make their food. Some invertebrates stay in one spot, while others fly, swim, float, crawl and burrow.

Types of Invertebrates

Eighty-five percent of invertebrates – some 923,000 species – are arthropods. Mollusks have approximately 100,000 distinct species. Some of the most common types of invertebrates are:

- Protozoans – single-celled organisms such as amoebas and paramecia.

- Annelids – earthworms, leeches.

- Echinoderms – starfish, sea urchins, sea cucumbers.

- Mollusks – snails, octopi, squid, snails, clams.

- Arthropods – insects, spiders, crustaceans such as shrimp, crabs, lobsters.

Example of Invertebrate Animals

Amoebas

Amoebas are single-celled organisms that are part of the simplest group of invertebrates. They have a cell membrane, DNA, a nucleus and organelles just like most cells in the human body. The difference is that amoebas can move using their pseudopodia, or "false feet." They also use pseudopodia to catch and eat food. Beneficial amoebas are found in soil where they help regulate bacterial populations and recycle nutrients. There are several species of ameba that cause disease, including Entameba hystolitica, which causes diarrhea.

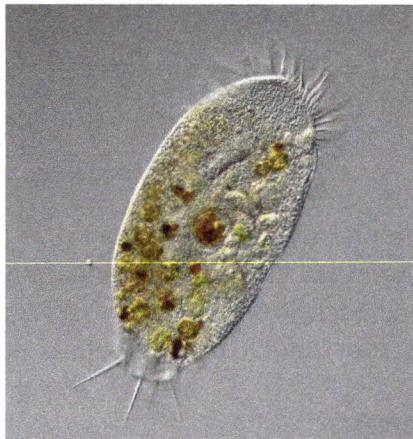

An amoeba looks like under a microscope.

Earthworms

Earthworms, also called angleworms, are very important in the world because their burrowing turns over and aerates the soil, provides soil drainage and mixes in organic material. They are one of more than 1,800 species of worms that live on land. Earthworms eat pieces of plants and animals in the soil. The bodies of worms are divided into segments; in certain species, some of the organs are duplicated in each segment. Earthworms can detect light and vibrations, but they are blind and cannot hear. Birds and other animals use earthworms as a food source.

Starfish

Starfish are also called sea stars because of the arms or "rays" they have sticking out of their bodies. Some starfish live deep in the oceans, while others live on the shore. There are over 1,600 species of starfish. Most have five arms, but one species is known to have 24 arms. Starfish have tube feet that work with hydraulic pressure to help them move. Snails, mussels and clams are common foods for starfish, which use their arms to guide food particles from the water into their mouths. These organisms have the unique ability to lose and regrow their arms.

Starfish.

Squids

Squids are decapods, meaning they have 10 arms or tentacles. They are different from octopuses, which have eight arms. Each tentacle has four rows of suckers that help them catch prey (usually fish) and attach themselves to surfaces. Squids live in the ocean, both near the shore and in the ocean depths. Their length ranges from about 3/4 of an inch to more than 65 feet. Some squids are luminous, which means they have light organs in their skin that let them change color to camouflage themselves, attract prey and communicate with other squids. Humans, some fishes and sperm whales eat squid.

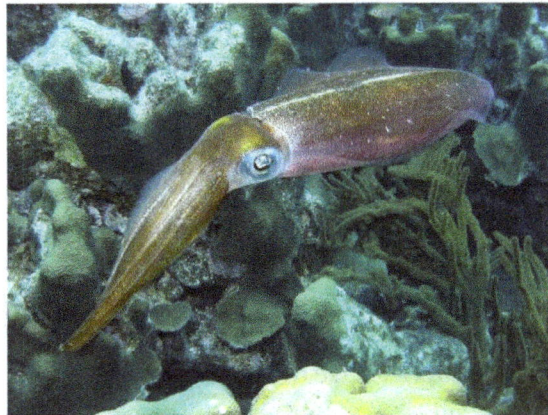
Reef squid.

Spiders

Spiders make their home on every continent except Antarctica. They are not insects because they have eight legs (insects have six legs). Experts think there are about 170,000 species of spiders in the world, but only around 39,000 are described and named. Some spiders live on land and others live in the water. The leg span of species ranges from 0.02 of an inch to 10 inches. Most spiders have eight eyes, and can sense light and dark. They also are very sensitive to vibration, which comes in handy when prey gets stuck in their web. Some spiders inject poison into their prey to kill it, while others prefer to wrap up their live catch and save it for later. Spiders can only ingest liquids; they bite holes in their prey and inject enzymes that dissolve the tissues so they can be sucked out.

ARTHROPODA

An arthropod is an invertebrate animal having an exoskeleton (external skeleton), a segmented body, and paired jointed appendages. Arthropods form the phylum Euarthropoda, which includes insects, arachnids, myriapods, and crustaceans. The term Arthropoda as originally proposed refers to a proposed grouping of Euarthropods and the phylum Onychophora. Arthropods are characterized by their jointed limbs and cuticle made of chitin, often mineralised with calcium carbonate. The arthropod body plan consists of segments, each with a pair of appendages. The rigid cuticle inhibits growth, so arthropods replace it periodically by moulting. Arthopods are bilaterally symmetrical and their body possesses an external skeleton. Some species have wings.

Their versatility has enabled them to become the most species-rich members of all ecological guilds in most environments. They have over a million described species, making up more than 80 percent of all described living animal species, some of which, unlike most other animals, are very successful in dry environments. Arthropods range in size from the microscopic crustacean Stygotantulus up to the Japanese spider crab.

Arthropods' primary internal cavity is a haemocoel, which accommodates their internal organs, and through which their haemolymph – analogue of blood – circulates; they have open circulatory systems. Like their exteriors, the internal organs of arthropods are generally built of repeated segments. Their nervous system is "ladder-like", with paired ventral nerve cords running through all segments and forming paired ganglia in each segment. Their heads are formed by fusion of varying numbers of segments, and their brains are formed by fusion of the ganglia of these segments and encircle the esophagus. The respiratory and excretory systems of arthropods vary, depending as much on their environment as on the subphylum to which they belong. Their vision relies on various combinations of compound eyes and pigment-pit ocelli: in most species the ocelli can only detect the direction from which light is coming, and the compound eyes are the main source of information, but the main eyes of spiders are ocelli that can form images and, in a few cases, can swivel to track prey. Arthropods also have a wide range of chemical and mechanical sensors, mostly based on modifications of the many bristles known as setae that project through their cuticles. Arthropods' methods of reproduction and development are diverse; all terrestrial species use internal fertilization, but this is often by indirect transfer of the sperm via an appendage or the ground, rather than by direct injection.

The evolutionary ancestry of arthropods dates back to the Cambrian period. The group is generally regarded as monophyletic, and many analyses support the placement of arthropods with cycloneuralians (or their constituent clades) in a superphylum Ecdysozoa. Overall, however, the basal relationships of animals are not yet well resolved. Likewise, the relationships between various arthropod groups are still actively debated. Aquatic species use either internal or external fertilization. Almost all arthropods lay eggs, but scorpions give birth to live young after the eggs have hatched inside the mother. Arthropod hatchlings vary from miniature adults to grubs and caterpillars that lack jointed limbs and eventually undergo a total metamorphosis to produce the adult form. The level of maternal care for hatchlings varies from nonexistent to the prolonged care provided by scorpions.

Arthropods contribute to the human food supply both directly as food, and more importantly indirectly as pollinators of crops. Some species are known to spread severe disease to humans, livestock, and crops.

Respiration and Circulation

Arthropods have open circulatory systems, although most have a few short, open-ended arteries. In chelicerates and crustaceans, the blood carries oxygen to the tissues, while hexapods use a separate system of tracheae. Many crustaceans, but few chelicerates and tracheates, use respiratory pigments to assist oxygen transport. The most common respiratory pigment in arthropods is copper-based hemocyanin; this is used by many crustaceans and a few centipedes. A few crustaceans and insects use iron-based hemoglobin, the respiratory pigment used by vertebrates. As with other invertebrates, the respiratory pigments of those arthropods that have them are generally dissolved in the blood and rarely enclosed in corpuscles as they are in vertebrates.

The heart is typically a muscular tube that runs just under the back and for most of the length of the hemocoel. It contracts in ripples that run from rear to front, pushing blood forwards. Sections not being squeezed by the heart muscle are expanded either by elastic ligaments or by small muscles, in either case connecting the heart to the body wall. Along the heart run a series of paired ostia, non-return valves that allow blood to enter the heart but prevent it from leaving before it reaches the front.

Arthropods have a wide variety of respiratory systems. Small species often do not have any, since their high ratio of surface area to volume enables simple diffusion through the body surface to supply enough oxygen. Crustacea usually have gills that are modified appendages. Many arachnids have book lungs. Tracheae, systems of branching tunnels that run from the openings in the body walls, deliver oxygen directly to individual cells in many insects, myriapods and arachnids.

Nervous System

Living arthropods have paired main nerve cords running along their bodies below the gut, and in each segment the cords form a pair of ganglia from which sensory and motor nerves run to other parts of the segment. Although the pairs of ganglia in each segment often appear physically fused, they are connected by commissures (relatively large bundles of nerves), which give arthropod nervous systems a characteristic "ladder-like" appearance. The brain is in the head, encircling and mainly above the esophagus. It consists of the fused ganglia of the acron and one or two of the foremost segments that form the head – a total of three pairs of ganglia in most arthropods, but only two in chelicerates, which do not have antennae or the ganglion connected to them. The ganglia of other head segments are often close to the brain and function as part of it. In insects these other head ganglia combine into a pair of subesophageal ganglia, under and behind the esophagus. Spiders take this process a step further, as all the segmental ganglia are incorporated into the subesophageal ganglia, which occupy most of the space in the cephalothorax (front "super-segment").

Excretory System

There are two different types of arthropod excretory systems. In aquatic arthropods, the end-product of biochemical reactions that metabolise nitrogen is ammonia, which is so toxic that it needs to be diluted as much as possible with water. The ammonia is then eliminated via any permeable membrane, mainly through the gills. All crustaceans use this system, and its high consumption of water may be responsible for the relative lack of success of crustaceans as land animals. Various groups of terrestrial arthropods have independently developed a different system: the end-product of nitrogen metabolism is uric acid, which can be excreted as dry material; the Malpighian tubule system filters the uric acid and other nitrogenous waste out of the blood in the hemocoel, and dumps these materials into the hindgut, from which they are expelled as feces. Most aquatic arthropods and some terrestrial ones also have organs called nephridia ("little kidneys"), which extract other wastes for excretion as urine.

Senses

Optical

The stiff cuticles of arthropods would block out information about the outside world, except that they are penetrated by many sensors or connections from sensors to the nervous system. In fact, arthropods have modified their cuticles into elaborate arrays of sensors. Various touch sensors, mostly setae, respond to different levels of force, from strong contact to very weak air currents. Chemical sensors provide equivalents of taste and smell, often by means of setae. Pressure sensors often take the form of membranes that function as eardrums, but are connected directly to nerves rather than to auditory ossicles. The antennae of most hexapods include sensor packages that monitor humidity, moisture and temperature.

Most arthropods have sophisticated visual systems that include one or more usually both of compound eyes and pigment-cup ocelli ("little eyes"). In most cases ocelli are only capable of detecting the direction from which light is coming, using the shadow cast by the walls of the cup. However, the main eyes of spiders are pigment-cup ocelli that are capable of forming images, and those of jumping spiders can rotate to track prey.

Compound eyes consist of fifteen to several thousand independent ommatidia, columns that are usually hexagonal in cross section. Each ommatidium is an independent sensor, with its own light-sensitive cells and often with its own lens and cornea. Compound eyes have a wide field of view, and can detect fast movement and, in some cases, the polarization of light. On the other hand, the relatively large size of ommatidia makes the images rather coarse, and compound eyes are shorter-sighted than those of birds and mammals – although this is not a severe disadvantage, as objects and events within 20 centimetres (7.9 in) are most important to most arthropods. Several arthropods have color vision, and that of some insects has been studied in detail; for example, the ommatidia of bees contain receptors for both green and ultra-violet.

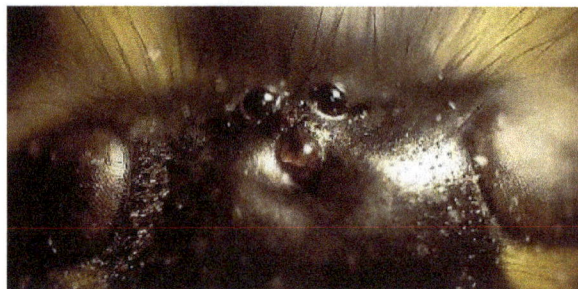

Head of a wasp with three ocelli (center), and compound eyes at the left and right.

Most arthropods lack balance and acceleration sensors, and rely on their eyes to tell them which way is up. The self-righting behavior of cockroaches is triggered when pressure sensors on the underside of the feet report no pressure. However, many malacostracan crustaceans have statocysts, which provide the same sort of information as the balance and motion sensors of the vertebrate inner ear.

The proprioceptors of arthropods, sensors that report the force exerted by muscles and the degree of bending in the body and joints, are well understood. However, little is known about what other internal sensors arthropods may have.

Classification

Arthropods belong to phylum Euarthropoda. The phylum is sometimes called Arthropoda, but strictly this term denotes a clade that also encompasses the phylum Onychophora.

Euarthropoda is typically subdivided into five subphyla, of which one is extinct:

- Trilobites are a group of formerly numerous marine animals that disappeared in the Permian–Triassic extinction event, though they were in decline prior to this killing blow, having been reduced to one order in the Late Devonian extinction.

- Chelicerates include horseshoe crabs, spiders, mites, scorpions and related organisms. They are characterised by the presence of chelicerae, appendages just above / in front of the mouth. Chelicerae appear in scorpions and horseshoe crabs as tiny claws that they use in feeding, but those of spiders have developed as fangs that inject venom.

- Myriapods comprise millipedes, centipedes, and their relatives and have many body segments, each segment bearing one or two pairs of legs (or in a few cases being legless). They are sometimes grouped with the hexapods.

- Crustaceans are primarily aquatic (a notable exception being woodlice) and are characterised by having biramous appendages. They include lobsters, crabs, barnacles, crayfish, shrimp and many others.

- Hexapods comprise insects and three small orders of insect-like animals with six thoracic legs. They are sometimes grouped with the myriapods, in a group called Uniramia, though genetic evidence tends to support a closer relationship between hexapods and crustaceans.

Aside from these major groups, there are also a number of fossil forms, mostly from the Early Cambrian, which are difficult to place, either from lack of obvious affinity to any of the main groups or from clear affinity to several of them. Marrella was the first one to be recognized as significantly different from the well-known groups.

The phylogeny of the major extant arthropod groups has been an area of considerable interest and dispute. Recent studies strongly suggest that Crustacea, as traditionally defined, is paraphyletic, with Hexapoda having evolved from within it, so that Crustacea and Hexapoda form a clade, Pancrustacea. The position of Myriapoda, Chelicerata and Pancrustacea remains unclear as of April 2012. In some studies, Myriapoda is grouped with Chelicerata (forming Myriochelata); in other studies, Myriapoda is grouped with Pancrustacea (forming Mandibulata), or Myriapoda may be sister to Chelicerata plus Pancrustacea.

The placement of the extinct trilobites is also a frequent subject of dispute. One of the newer hypotheses is that the chelicerae have originated from the same pair of appendages that evolved into antennae in the ancestors of Mandibulata, which would place trilobites, which had antennae, closer to Mandibulata than Chelicerata.

Since the International Code of Zoological Nomenclature recognises no priority above the rank of family, many of the higher-level groups can be referred to by a variety of different names.

ARACHNID

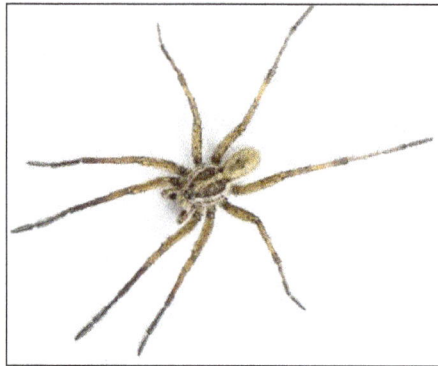

Arachnid (class Arachnida), is the member of the arthropod group that includes spiders, daddy longlegs, scorpions, and (in the subclass Acari) the mites and ticks, as well as lesser-known subgroups. Only a few species are of economic importance—for example, the mites and ticks, which transmit diseases to humans, other animals, and plants.

Features

Body and Appendages

Arachnids range in size from tiny mites that measure 0.08 mm (0.003 inch) to the enormous scorpion Hadogenes troglodytes of Africa, which may be 21 cm (8 inches) or more in length. In appearance, they vary from short-legged, round-bodied mites and pincer-equipped scorpions with curled tails to delicate, long-legged daddy longlegs and robust, hairy tarantulas.

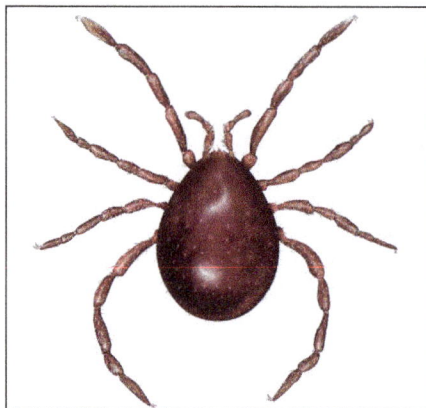

Acari (Holothyrus legendrei).

Like all arthropods, arachnids have segmented bodies, tough exoskeletons, and jointed append-ages. Most are predatory. Arachnids lack jaws and, with only a few exceptions, inject digestive fluids into their prey before sucking its liquefied remains into their mouths. Except among daddy longlegs and the mites and ticks, in which the entire body forms a single region, the arachnid body is divided into two distinct regions: the cephalothorax, or prosoma, and the abdomen, or opistho-soma. The sternites (ventral plates) of the lower surface of the body show more variation than do the tergites (dorsal plates). The arachnids have simple (as opposed to compound) eyes.

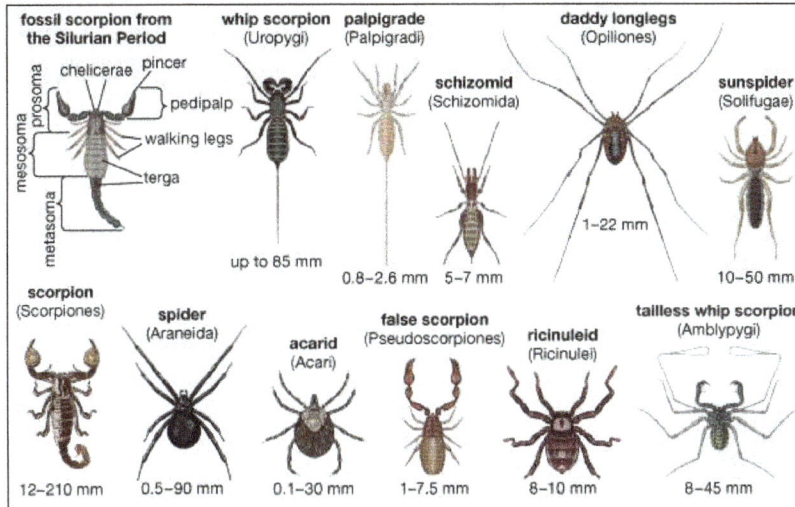

Diversity of arachnids.

The cephalothorax is covered dorsally with a rigid cover (the carapace) and has six pairs of ap-pendages, the first of which are the chelicerae, the only appendages that are in front of the mouth. In many forms they are chelate, or pincerlike, and are used to hold and crush prey. Among spiders the basal segment of the chelicerae contains venom sacs, and the second segment, the fang, in-jects venom. The pedipalps, or palps, which in arachnids function as an organ of touch, constitute the second pair of appendages. In spiders and daddy longlegs the pedipalps are elongated leglike structures, whereas in scorpions they are large chelate, prehensile organs. Among spiders the pedi-palps are highly modified as secondary sexual organs. The basal segment is sometimes modified for crushing or cutting food. The remaining four pairs of appendages are walking legs, though the first of these pairs serves as tactile organs among the tailless whip scorpions (order Amblypygi); it is the second pair that functions as such among the daddy longlegs. Among the spiderlike ricinu-leids (order Ricinulei), special copulatory organs are located on the third pair of legs. Some mites, particularly immature individuals, have only two or three pairs of legs.

A ricinuleid (order Ricinulei).

In many arachnids the cephalothorax and abdomen are broadly joined, while in others (such as spiders) they are joined by a narrow stalklike pedicel. The abdomen is composed of a maximum (in scorpions) of 13 segments, or somites. The first of these may be present only in the embryo and absent in the adult. In some orders a mesosoma consisting of seven segments and metasoma of five may be distinguished, while in others a few posterior segments may form a postabdomen (pygidium). In general, except for the spinnerets of the spiders, the abdomen has no appendages. In some groups it is elongated and distinctly segmented; in others it may be shortened, with indistinct segmentation. Postanal structures vary in both appearance and function. The scorpions have a short stinger with a swollen base enclosing a poison gland, and the whip scorpions (order Uropygi) and micro whip scorpions (order Palpigradi) have long whiplike structures of unknown function.

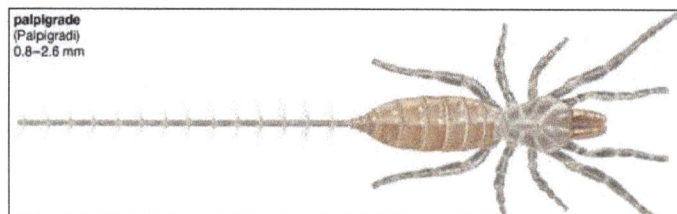

palpigrade
(Palpigradi)
0.8–2.6 mm

A palpigrade (order Palpigradi).

Distribution and Abundance

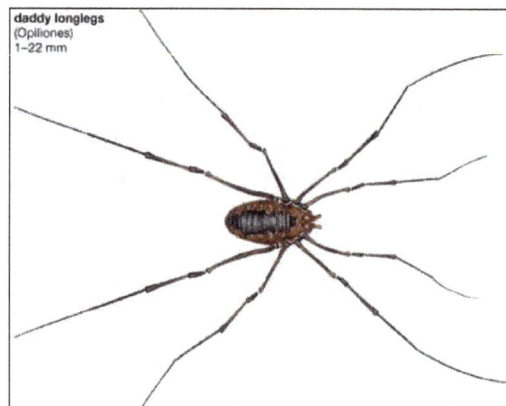

daddy longlegs
(Opiliones)
1–22 mm

Daddy longlegs, or harvestman (order Opiliones).

With the exception of a few groups that have become aquatic, arachnids are terrestrial predators. Spiders (order Araneida), daddy longlegs (or harvestmen; order Opiliones), false scorpions (order Pseudoscorpiones), and mites and ticks (subclass Acari) are nearly worldwide in distribution. Scorpions (order Scorpiones), sunspiders (or wind scorpions; order Solpugida), tailless whip scorpions (order Amblypygi), and micro whip scorpions (or vinegarroons; order Uropygi) are widespread within the tropical and subtropical areas of the world, only occasionally being encountered in temperate areas. Of more sporadic distribution but more common in tropical areas are the sunspiders, the schizomids (order Schizomida), and the ricinuleids (order Ricinulei). In temperate areas mature spiders and daddy longlegs are particularly conspicuous during early autumn, though they are abundant throughout the year. Most arachnids, however, are seldom observed, for they inhabit leaf mold and litter or soil. Most abundant of the arachnids are the ticks and mites, found in soil, in fresh and marine waters, and as parasites of animals, including humans.

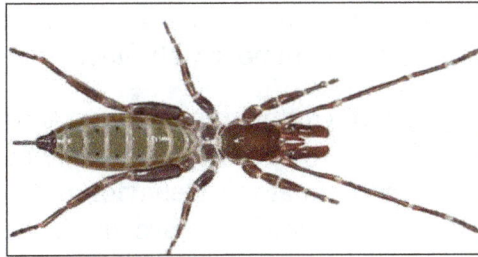
A schizomid (order Schizomida).

The numbers and predaceous habits of arachnids make them important to humans. Free-living mites play an important role in the conversion of leaf mold to humus. Many mites are parasitic, and many ticks are intermediate hosts for organisms that cause serious diseases. Though all spiders possess poison that can be utilized for subduing prey, only a few have a poison sufficiently powerful to affect humans. A bite of the black widow spider (Latrodectus mactans) may result in discomfort or serious illness, whereas that of the brown recluse spider (Loxosceles reclusa) may result in a severe local reaction, including tissue death. The sting of some scorpions may cause a severe reaction and even death.

Black widow spider (Latrodectus mactans).

Classification

Distinguishing Taxonomic Features

In classifying arachnids, taxonomists rely mostly upon external structures, including such features as general body form, the degree of visible external segmentation, structural modifications of the prosoma and opisthosoma, characteristics of appendages, and special structures involved in sperm transfer. Internal anatomical features, developmental traits, and serological characteristics are used to a limited extent. However, as more information becomes available at the molecular level, traits such as these could play a more important role in arachnid classification.

MOLLUSCA

Mollusca is the second-largest phylum of invertebrate animals. The members are known as molluscs or mollusks. Around 85,000 extant species of molluscs are recognized. The number of fossil species is estimated between 60,000 and 100,000 additional species. The proportion of undescribed species is very high. Many taxa remain poorly studied.

Molluscs are the largest marine phylum, comprising about 23% of all the named marine organisms. Numerous molluscs also live in freshwater and terrestrial habitats. They are highly diverse, not just in size and anatomical structure, but also in Behavior and habitat. The phylum is typically divided into 8 or 9 taxonomic classes, of which two are entirely extinct. Cephalopod molluscs, such as squid, cuttlefish, and octopuses, are among the most neurologically advanced of all invertebrates—and either the giant squid or the colossal squid is the largest known invertebrate species. The gastropods (snails and slugs) are by far the most numerous molluscs and account for 80% of the total classified species.

The three most universal features defining modern molluscs are a mantle with a significant cavity used for breathing and excretion, the presence of a radula (except for bivalves), and the structure of the nervous system. Other than these common elements, molluscs express great morphological diversity, so many textbooks base their descriptions on a "hypothetical ancestral mollusc". This has a single, "limpet-like" shell on top, which is made of proteins and chitin reinforced with calcium carbonate, and is secreted by a mantle covering the whole upper surface. The underside of the animal consists of a single muscular "foot". Although molluscs are coelomates, the coelom tends to be small. The main body cavity is a hemocoel through which blood circulates; as such, their circulatory systems are mainly open. The "generalized" mollusc's feeding system consists of a rasping "tongue", the radula, and a complex digestive system in which exuded mucus and microscopic, muscle-powered "hairs" called cilia play various important roles. The generalized mollusc has two paired nerve cords, or three in bivalves. The brain, in species that have one, encircles the esophagus. Most molluscs have eyes, and all have sensors to detect chemicals, vibrations, and touch. The simplest type of molluscan reproductive system relies on external fertilization, but more complex variations occur. All produce eggs, from which may emerge trochophore larvae, more complex veliger larvae, or miniature adults. The coelomic cavity is reduced. They have an open circulatory system and kidney-like organs for excretion.

Cornu aspersum (formerly Helix aspersa) – a common land snail.

Good evidence exists for the appearance of gastropods, cephalopods, and bivalves in the Cambrian period, 541 to 485.4 million years ago. However, the evolutionary history both of molluscs' emergence from the ancestral Lophotrochozoa and of their diversification into the well-known living and fossil forms are still subjects of vigorous debate among scientists.

Molluscs have been and still are an important food source for anatomically modern humans. A risk of food poisoning exists from toxins that can accumulate in certain molluscs under specific conditions, however, and because of this, many countries have regulations to reduce this risk. Molluscs have, for centuries, also been the source of important luxury goods, notably pearls, mother of pearl, Tyrian purple dye, and sea silk. Their shells have also been used as money in some preindustrial societies.

Mollusc species can also represent hazards or pests for human activities. The bite of the blue-ringed octopus is often fatal, and that of Octopus apollyon causes inflammation that can last over a month. Stings from a few species of large tropical cone shells can also kill, but their sophisticated, though easily produced, venoms have become important tools in neurological research. Schistosomiasis (also known as bilharzia, bilharziosis, or snail fever) is transmitted to humans by water snail hosts, and affects about 200 million people. Snails and slugs can also be serious agricultural pests, and accidental or deliberate introduction of some snail species into new environments has seriously damaged some ecosystems.

Hypothetical Ancestral Mollusc

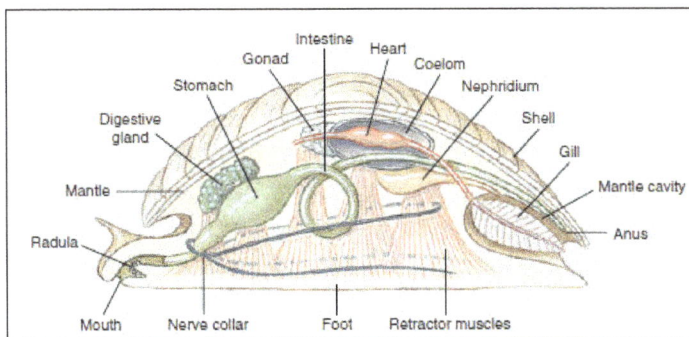

Anatomical diagram of a hypothetical ancestral mollusc.

Because of the great range of anatomical diversity among molluscs, many textbooks start the subject of molluscan anatomy by describing what is called an archi-mollusc, hypothetical generalized mollusc, or hypothetical ancestral mollusc (HAM) to illustrate the most common features found within the phylum. The depiction is visually rather similar to modern monoplacophorans.

The generalized mollusc is bilaterally symmetrical and has a single, "limpet-like" shell on top. The shell is secreted by a mantle covering the upper surface. The underside consists of a single muscular "foot". The visceral mass, or visceropallium, is the soft, nonmuscular metabolic region of the mollusc. It contains the body organs.

Mantle and Mantle Cavity

The mantle cavity, a fold in the mantle, encloses a significant amount of space. It is lined with epidermis, and is exposed, according to habitat, to sea, fresh water or air. The cavity was at the rear in the earliest molluscs, but its position now varies from group to group. The anus, a pair of osphradia (chemical sensors) in the incoming "lane", the hindmost pair of gills and the exit openings of the nephridia ("kidneys") and gonads (reproductive organs) are in the mantle cavity. The whole soft body of bivalves lies within an enlarged mantle cavity.

Shell

The mantle edge secretes a shell (secondarily absent in a number of taxonomic groups, such as the nudibranchs) that consists of mainly chitin and conchiolin (a protein hardened with calcium carbonate), except the outermost layer, which in almost all cases is all conchiolin Molluscs never use phosphate to construct their hard parts, with the questionable exception of Cobcrephora. While most mollusc shells are composed mainly of aragonite, those gastropods that lay eggs with a hard shell use calcite (sometimes with traces of aragonite) to construct the eggshells.

The shell consists of three layers: the outer layer (the periostracum) made of organic matter, a middle layer made of columnar calcite, and an inner layer consisting of laminated calcite, often nacreous.

In some forms the shell contains openings. In abalones there are holes in the shell used for respiration and the release of egg and sperm, in the nautilus a string of tissue called the siphuncle goes through all the chambers, and the eight plates that make up the shell of chitons are penetrated with living tissue with nerves and sensory structures.

Foot

The underside consists of a muscular foot, which has adapted to different purposes in different classes. The foot carries a pair of statocysts, which act as balance sensors. In gastropods, it secretes mucus as a lubricant to aid movement. In forms having only a top shell, such as limpets, the foot acts as a sucker attaching the animal to a hard surface, and the vertical muscles clamp the shell down over it; in other molluscs, the vertical muscles pull the foot and other exposed soft parts into the shell. In bivalves, the foot is adapted for burrowing into the sediment in cephalopods it is used for jet propulsion, and the tentacles and arms are derived from the foot.

Circulatory System

Most molluscs' circulatory systems are mainly open. Although molluscs are coelomates, their coeloms are reduced to fairly small spaces enclosing the heart and gonads. The main body cavity is a hemocoel through which blood and coelomic fluid circulate and which encloses most of the other internal organs. These hemocoelic spaces act as an efficient hydrostatic skeleton. The blood of these molluscs contains the respiratory pigment hemocyanin as an oxygen-carrier. The heart consists of one or more pairs of atria (auricles), which receive oxygenated blood from the gills and pump it to the ventricle, which pumps it into the aorta (main artery), which is fairly short and opens into the hemocoel. The atria of the heart also function as part of the excretory system by filtering waste products out of the blood and dumping it into the coelom as urine. A pair of nephridia ("little kidneys") to the rear of and connected to the coelom extracts any re-usable materials from the urine and dumps additional waste products into it, and then ejects it via tubes that discharge into the mantle cavity.

Exceptions to the above are the molluscs Planorbidae or ram's horn snails, which are air-breathing snails that use iron-based hemoglobin instead of the copper-based hemocyanin to carry oxygen through their blood.

Respiration

Most molluscs have only one pair of gills, or even only a singular gill. Generally, the gills are rather like feathers in shape, although some species have gills with filaments on only one side. They divide the mantle cavity so water enters near the bottom and exits near the top. Their filaments have three kinds of cilia, one of which drives the water current through the mantle cavity, while the other two help to keep the gills clean. If the osphradia detect noxious chemicals or possibly sediment entering the mantle cavity, the gills' cilia may stop beating until the unwelcome intrusions have ceased. Each gill has an incoming blood vessel connected to the hemocoel and an outgoing one to the heart.

Eating, Digestion and Excretion

Members of the mollusc family use intracellular digestion to function. Most molluscs have muscular mouths with radulae, "tongues", bearing many rows of chitinous teeth, which are replaced from the rear as they wear out. The radula primarily functions to scrape bacteria and algae off rocks, and is associated with the odontophore, a cartilaginous supporting organ. The radula is unique to the molluscs and has no equivalent in any other animal.

Molluscs' mouths also contain glands that secrete slimy mucus, to which the food sticks. Beating cilia (tiny "hairs") drive the mucus towards the stomach, so the mucus forms a long string called a "food string".

At the tapered rear end of the stomach and projecting slightly into the hindgut is the prostyle, a backward-pointing cone of feces and mucus, which is rotated by further cilia so it acts as a bobbin, winding the mucus string onto itself. Before the mucus string reaches the prostyle, the acidity of the stomach makes the mucus less sticky and frees particles from it.

The particles are sorted by yet another group of cilia, which send the smaller particles, mainly minerals, to the prostyle so eventually they are excreted, while the larger ones, mainly food, are sent to the stomach's cecum (a pouch with no other exit) to be digested. The sorting process is by no means perfect.

Periodically, circular muscles at the hindgut's entrance pinch off and excrete a piece of the prostyle, preventing the prostyle from growing too large. The anus, in the part of the mantle cavity, is swept by the outgoing "lane" of the current created by the gills. Carnivorous molluscs usually have simpler digestive systems.

As the head has largely disappeared in bivalves, the mouth has been equipped with labial palps (two on each side of the mouth) to collect the detritus from its mucus.

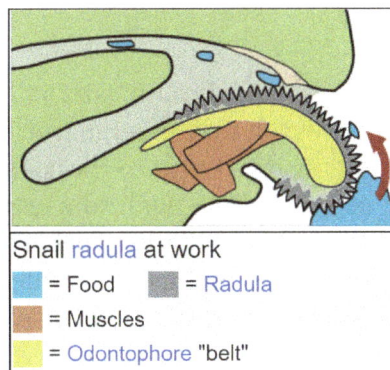

Snail radula at work
■ = Food   ■ = Radula
■ = Muscles
■ = Odontophore "belt"

Nervous System

The cephalic molluscs have two pairs of main nerve cords organized around a number of paired ganglia, the visceral cords serving the internal organs and the pedal ones serving the foot. Most pairs of corresponding ganglia on both sides of the body are linked by commissures (relatively large bundles of nerves). The ganglia above the gut are the cerebral, the pleural, and the visceral, which are located above the esophagus (gullet). The pedal ganglia, which control the foot, are below the esophagus and their commissure and connectives to the cerebral and pleural ganglia surround the esophagus in a circumesophageal nerve ring or nerve collar.

The acephalic molluscs (i.e., bivalves) also have this ring but it is less obvious and less important. The bivalves have only three pairs of ganglia— cerebral, pedal, and visceral— with the visceral as the largest and most important of the three functioning as the principal center of "thinking". Some such as the scallops have eyes around the edges of their shells which connect to a pair of looped nerves and which provide the ability to distinguish between light and shadow.

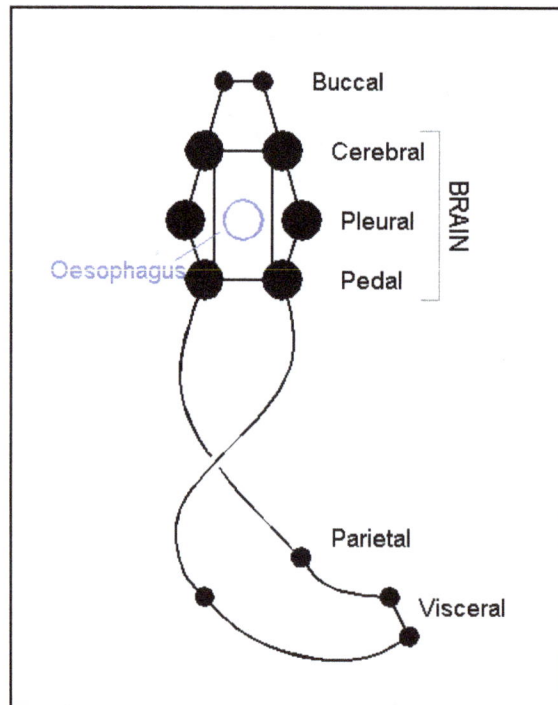

Simplified diagram of the mollusc nervous system.

Reproduction

The simplest molluscan reproductive system relies on external fertilization, but with more complex variations. All produce eggs, from which may emerge trochophore larvae, more complex veliger larvae, or miniature adults. Two gonads sit next to the coelom, a small cavity that surrounds the heart, into which they shed ova or sperm. The nephridia extract the gametes from the coelom and emit them into the mantle cavity. Molluscs that use such a system remain of one sex all their lives and rely on external fertilization. Some molluscs use internal fertilization and/or are hermaphrodites, functioning as both sexes; both of these methods require more complex reproductive systems.

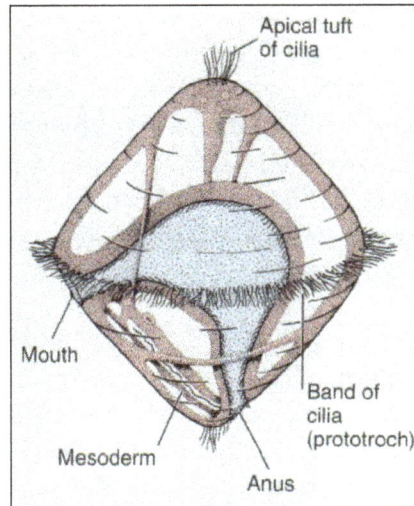

Trochophore larva.

The most basic molluscan larva is a trochophore, which is planktonic and feeds on floating food particles by using the two bands of cilia around its "equator" to sweep food into the mouth, which uses more cilia to drive them into the stomach, which uses further cilia to expel undigested remains through the anus. New tissue grows in the bands of mesoderm in the interior, so the apical tuft and anus are pushed further apart as the animal grows. The trochophore stage is often succeeded by a veliger stage in which the prototroch, the "equatorial" band of cilia nearest the apical tuft, develops into the velum ("veil"), a pair of cilia-bearing lobes with which the larva swims. Eventually, the larva sinks to the seafloor and metamorphoses into the adult form. While metamorphosis is the usual state in molluscs, the cephalopods differ in exhibiting direct development: the hatchling is a 'miniaturized' form of the adult.

ANNELID

The annelids also known as the ringed worms or segmented worms, are a large phylum, with over 22,000 extant species including ragworms, earthworms, and leeches. The species exist in and have adapted to various ecologies – some in marine environments as distinct as tidal zones and hydro-thermal vents, others in fresh water, and yet others in moist terrestrial environments.

The annelids are bilaterally symmetrical, triploblastic, coelomate, invertebrate organisms. They also have parapodia for locomotion. Most textbooks still use the traditional division into poly-chaetes (almost all marine), oligochaetes (which include earthworms) and leech-like species. Cladistic research since 1997 has radically changed this scheme, viewing leeches as a sub-group of oligochaetes and oligochaetes as a sub-group of polychaetes. In addition, the Pogonophora, Echiura and Sipuncula, previously regarded as separate phyla, are now regarded as sub-groups of polychaetes. Annelids are considered members of the Lophotrochozoa, a "super-phylum" of pro-tostomes that also includes molluscs, brachiopods, flatworms and nemerteans.

The basic annelid form consists of multiple segments. Each segment has the same sets of organs and, in most polychates, has a pair of parapodia that many species use for locomotion. Septa separate the segments of many species, but are poorly defined or absent in others, and Echiura and Sipuncula show no obvious signs of segmentation. In species with well-developed septa, the blood circulates entirely within blood vessels, and the vessels in segments near the front ends of these species are often built up with muscles that act as hearts. The septa of such species also enable them to change the shapes of individual segments, which facilitates movement by peristalsis ("ripples" that pass along the body) or by undulations that improve the effectiveness of the parapodia. In species with incomplete septa or none, the blood circulates through the main body cavity without any kind of pump, and there is a wide range of locomotory techniques – some burrowing species turn their pharynges inside out to drag themselves through the sediment.

Earthworms are oligochaetes that support terrestrial food chains both as prey and in some regions are important in aeration and enriching of soil. The burrowing of marine polychaetes, which may constitute up to a third of all species in near-shore environments, encourages the development of ecosystems by enabling water and oxygen to penetrate the sea floor. In addition to improving soil fertility, annelids serve humans as food and as bait. Scientists observe annelids to monitor the quality of marine and fresh water. Although blood-letting is used less frequently by doctors, some leech species are regarded as endangered species because they have been over-harvested for this purpose in the last few centuries. Ragworms' jaws are now being studied by engineers as they offer an exceptional combination of lightness and strength.

Since annelids are soft-bodied, their fossils are rare – mostly jaws and the mineralized tubes that some of the species secreted. Although some late Ediacaran fossils may represent annelids, the oldest known fossil that is identified with confidence comes from about 518 million years ago in the early Cambrian period. Fossils of most modern mobile polychaete groups appeared by the end of the Carboniferous, about 299 million years ago. Palaeontologists disagree about whether some body fossils from the mid Ordovician, about 472 to 461 million years ago, are the remains of oligochaetes, and the earliest indisputable fossils of the group appear in the Tertiary period, which began 66 million years ago.

Classification and Diversity

There are over 22,000 living annelid species, ranging in size from microscopic to the Australian giant Gippsland earthworm and Amynthas mekongianus, which can both grow up to 3 metres (9.8 ft) long. Although research since 1997 has radically changed scientists' views about the evolutionary family tree of the annelids, most textbooks use the traditional classification into the following sub-groups:

- Polychaetes (about 12,000 species). As their name suggests, they have multiple chetae ("hairs") per segment. Polychaetes have parapodia that function as limbs, and nuchal organs that are thought to be chemosensors. Most are marine animals, although a few species live in fresh water and even fewer on land.

- Clitellates (about 10,000 species). These have few or no chetae per segment, and no nuchal organs or parapodia. However, they have a unique reproductive organ, the ring-shaped clitellum ("pack saddle") around their bodies, which produces a cocoon that stores and nourishes fertilized eggs until they hatch or, in moniligastrids, yolky eggs that provide nutrition for the embryos. The clitellates are sub-divided into:

    ○ Oligochaetes ("with few hairs"), which includes earthworms. Oligochaetes have a sticky pad in the roof of the mouth. Most are burrowers that feed on wholly or partly decomposed organic materials.

○   Hirudinea, whose name means "leech-shaped" and whose best known members are leeches. Marine species are mostly blood-sucking parasites, mainly on fish, while most freshwater species are predators. They have suckers at both ends of their bodies, and use these to move rather like inchworms.

The Archiannelida, minute annelids that live in the spaces between grains of marine sediment, were treated as a separate class because of their simple body structure, but are now regarded as polychaetes. Some other groups of animals have been classified in various ways, but are now widely regarded as annelids:

- Pogonophora / Siboglinidae were first discovered in 1914, and their lack of a recognizable gut made it difficult to classify them. They have been classified as a separate phylum, Pogonophora, or as two phyla, Pogonophora and Vestimentifera. More recently they have been re-classified as a family, Siboglinidae, within the polychaetes.

- The Echiura have a checkered taxonomic history: in the 19th century they were assigned to the phylum "Gephyrea", which is now empty as its members have been assigned to other phyla; the Echiura were next regarded as annelids until the 1940s, when they were classified as a phylum in their own right; but a molecular phylogenetics analysis in 1997 concluded that echiurans are annelids.

- Myzostomida live on crinoids and other echinoderms, mainly as parasites. In the past they have been regarded as close relatives of the trematode flatworms or of the tardigrades, but in 1998 it was suggested that they are a sub-group of polychaetes. However, another analysis in 2002 suggested that myzostomids are more closely related to flatworms or to rotifers and acanthocephales.

- Sipuncula was originally classified as annelids, despite the complete lack of segmentation, bristles and other annelid characters. The phylum Sipuncula was later allied with the Mollusca, mostly on the basis of developmental and larval characters. Phylogenetic analyses based on 79 ribosomal proteins indicated a position of Sipuncula within Annelida. Subsequent analysis of the mitochondrion's DNA has confirmed their close relationship to the Myzostomida and Annelida (including echiurans and pogonophorans). It has also been shown that a rudimentary neural segmentation similar to that of annelids occurs in the early larval stage, even if these traits are absent in the adults.

No single feature distinguishes Annelids from other invertebrate phyla, but they have a distinctive combination of features. Their bodies are long, with segments that are divided externally by shallow ring-like constrictions called annuli and internally by septa ("partitions") at the same points, although in some species the septa are incomplete and in a few cases missing. Most of the segments contain the same sets of organs, although sharing a common gut, circulatory system and nervous system makes them inter-dependent. Their bodies are covered by a cuticle (outer covering) that does not contain cells but is secreted by cells in the skin underneath, is made of tough but flexible collagen and does not molt – on the other hand arthropods' cuticles are made of the more rigid α-chitin, and molt until the arthropods reach their full size. Most annelids have closed circulatory systems, where the blood makes its entire circuit via blood vessels.

NEMATODE

The nematodes or roundworms constitute the phylum Nematoda (also called Nemathelminthes). They are a diverse animal phylum inhabiting a broad range of environments. Taxonomically, they are classified along with insects and other moulting animals in the clade Ecdysozoa, and unlike flatworms, have tubular digestive systems with openings at both ends.

Nematode species can be difficult to distinguish from one another. Consequently, estimates of the number of nematode species described to date vary by author and may change rapidly over time. A 2013 survey of animal biodiversity published in the mega journal Zootaxa puts this figure at over 25,000. Estimates of the total number of extant species are subject to even greater variation.

Nematodes have successfully adapted to nearly every ecosystem: from marine (salt) to fresh water, soils, from the polar regions to the tropics, as well as the highest to the lowest of elevations. They are ubiquitous in freshwater, marine, and terrestrial environments, where they often outnumber other animals in both individual and species counts, and are found in locations as diverse as mountains, deserts, and oceanic trenches. They are found in every part of the earth's lithosphere, even at great depths, 0.9–3.6 km (3,000–12,000 ft) below the surface of the Earth in gold mines in South Africa. They represent 90% of all animals on the ocean floor. Their numerical dominance, often exceeding a million individuals per square meter and accounting for about 80% of all individual animals on earth, their diversity of lifecycles, and their presence at various trophic levels point to an important role in many ecosystems. They have been shown to play crucial roles in polar ecosystem. The roughly 2,271 genera are placed in 256 families. The many parasitic forms include pathogens in most plants and animals. A third of the genera occur as parasites of vertebrates; about 35 nematode species occur in humans.

Anatomy

Nematodes are very small, slender worms: typically about 5 to 100 µm thick, and 0.1 to 2.5 mm long. The smallest nematodes are microscopic, while free-living species can reach as much as 5 cm (2 in), and some parasitic species are larger still, reaching over 1 m (3 ft) in length. The body is often ornamented with ridges, rings, bristles, or other distinctive structures.

The head of a nematode is relatively distinct. Whereas the rest of the body is bilaterally symmetrical, the head is radially symmetrical, with sensory bristles and, in many cases, solid 'headshields' radiating outwards around the mouth. The mouth has either three or six lips, which often bear a series of teeth on their inner edges. An adhesive 'caudal gland' is often found at the tip of the tail.

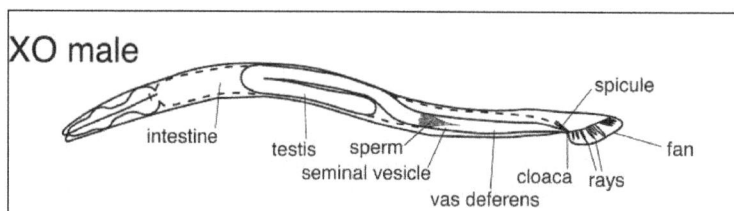

Internal anatomy of a male C. elegans nematode.

The epidermis is either a syncytium or a single layer of cells, and is covered by a thick collagenous cuticle. The cuticle is often of complex structure, and may have two or three distinct layers. Underneath the epidermis lies a layer of longitudinal muscle cells. The relatively rigid cuticle works with the muscles to create a hydroskeleton, as nematodes lack circumferential muscles. Projections run from the inner surface of muscle cells towards the nerve cords; this is a unique arrangement in the animal kingdom, in which nerve cells normally extend fibres into the muscles rather than vice versa.

Digestive System

The oral cavity is lined with cuticle, which is often strengthened with structures, such as ridges, and especially in carnivorous species, may bear a number of teeth. The mouth often includes a sharp stylet, which the animal can thrust into its prey. In some species, the stylet is hollow, and can be used to suck liquids from plants or animals.

The oral cavity opens into a muscular, sucking pharynx, also lined with cuticle. Digestive glands are found in this region of the gut, producing enzymes that start to break down the food. In stylet-bearing species, these may even be injected into the prey.

No stomach is present, with the pharynx connecting directly to a muscleless intestine that forms the main length of the gut. This produces further enzymes, and also absorbs nutrients through its single-cell-thick lining. The last portion of the intestine is lined by cuticle, forming a rectum, which expels waste through the anus just below and in front of the tip of the tail. Movement of food through the digestive system is the result of body movements of the worm. The intestine has valves or sphincters at either end to help control the movement of food through the body.

Excretory System

Nitrogenous waste is excreted in the form of ammonia through the body wall, and is not associated with any specific organs. However, the structures for excreting salt to maintain osmoregulation are typically more complex.

In many marine nematodes, one or two unicellular 'renette glands' excrete salt through a pore on the underside of the animal, close to the pharynx. In most other nematodes, these specialised cells have been replaced by an organ consisting of two parallel ducts connected by a single transverse duct. This transverse duct opens into a common canal that runs to the excretory pore.

Nervous System

Four peripheral nerves run the length of the body on the dorsal, ventral, and lateral surfaces. Each nerve lies within a cord of connective tissue lying beneath the cuticle and between the muscle cells. The ventral nerve is the largest, and has a double structure forward of the excretory pore. The dorsal nerve is responsible for motor control, while the lateral nerves are sensory, and the ventral combines both functions.

The nervous system is also the only place in the nematode body that contains cilia, which are all nonmotile and with a sensory function.

At the anterior end of the animal, the nerves branch from a dense, circular nerve (nerve ring) round surrounding the pharynx, and serving as the brain. Smaller nerves run forward from the ring to supply the sensory organs of the head.

The bodies of nematodes are covered in numerous sensory bristles and papillae that together provide a sense of touch. Behind the sensory bristles on the head lie two small pits, or 'amphids'. These are well supplied with nerve cells, and are probably chemoreception organs. A few aquatic nematodes possess what appear to be pigmented eye-spots, but whether or not these are actually sensory in nature is unclear.

References

- Chapman, A.D. (2009). Numbers of Living Species in Australia and the World, 2nd edition. Australian Biological Resources Study, Canberra. Retrieved 2010-01-12. ISBN 978-0-642-56860-1 (printed); ISBN 978-0-642-56861-8 (online)

- Invertebrate: biologydictionary.net, Retrieved 24 July, 2019

- J. Ortega-Hernández (February 2016), "Making sense of 'lower' and 'upper' stem-group Euarthropoda, with comments on the strict use of the name Arthropoda von Siebold, 1848", Biological Reviews, 91 (1): 255–273, doi:10.1111/brv.12168, PMID 25528950

- Respiration, arachnid: britannica.com, Retrieved 25 August, 2019

- Williams, D.M. (April 21, 2001), "Largest", Book of Insect Records, University of Florida, archived from the original on July 18, 2011, retrieved 2009-06-10

5

Animal Physiology and Anatomy

The study of the life-supporting properties, processes and functions of animals is known as animal physiology. The branch of biology which studies the structure of animals as well as their parts is known as animal anatomy. The topics elaborated in this chapter will help in gaining a better perspective about the major areas of study within animal physiology and anatomy.

ANIMAL ANATOMY

The study of the structure of living things is called anatomy. All animals are made up of cells, some of which are specialized to carry out different functions. Simple animals, such as sponges, are made up of only a few types of cell. In more complex animals, cells are organized into tissues, such as muscles and nerves that are necessary for movement. Tissues can form organs, such as the heart, which is used to pump blood around the circulatory system.

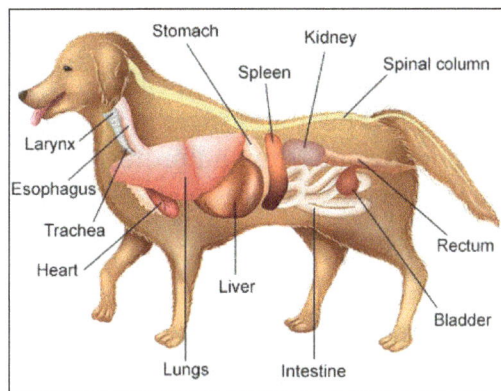

All animals with backbones have an internal framework of support, called an endoskeleton. Bony skeletons, such as that of the squirrel, are light to aid movement. When an animal is young, bones in the skeleton can grow in length. Some bones protect vital organs, while limb bones provide anchorage for muscles.

All animals need oxygen to survive. Simple animals exchange gases over the surface of their bodies. Insects, such as caterpillars, have openings along their bodies, called spiracles, through which air passes. Animals with lungs, such as birds and mammals, actively breathe.

Perfect Symmetry

Most animals are bilaterally symmetrical. If a penguin were cut in half from head to toe, the two halves would be mirror images of each other. Other animals, such as sea anemones, are radially symmetrical. They have no head or tail and can be cut into identical halves along many lines. Of the two types, animals that are bilaterally symmetrical tend to move more quickly and precisely.

Shark Anatomy

Like all fish, sharks have a backbone, breathe through gills, manoeuvre using fins, and are ecto-thermic (cold-blooded). A shark's anatomy also bears the hallmarks of a predatory fish. They have a streamlined, torpedo-shaped body that allows them to cut easily through water to chase prey. They also have powerful jaws and sharp teeth.

Exoskeleton

Like all arthropods, lobsters have a hard outer casing, called an exoskeleton, made up of plates formed from a substance called chitin. The plates meet at flexible points, such as the leg joints. This exoskeleton provides anchorage for muscles and protection from predators. It also provides support for movement on land and prevents excess water loss.

Cells

Animal cells are typically just 0.02 mm (1/1,250 in) across. Although they can be extremely var-ied, they share common features. Cells are surrounded by a skin called a membrane and contain a jelly-like fluid called cytoplasm. All the processes needed for life, such as producing energy from food, removing waste, and growth take place inside cells.

Cell Components

Inside an animal cell, the cytoplasm contains structures called organelles that have a variety of functions, from storing vital substances to destroying bacteria. The most important organelle is called the nucleus, which carries genetic information, controlling how the cell behaves. Another organelle, the mitochondrion, produces energy from food.

Circulatory System

The circulatory system carries blood around an animal's body, providing nourishment and oxygen to cells. In some animals it is open, in others it is closed. In an open system, blood flows freely around the body. In a closed system, blood is confined to a network of vessels. The circulatory system also helps distribute heat around the body.

Many land animals, such as reptiles, are ectothermic – they rely on the Sun's heat to raise their body temperature to a level that allows them to be active. Birds and mammals are endothermic – they produce their own heat and maintain a constant body temperature.

Movement

All animals are mobile for at least some part of their lives because they need to find food. Most movement is controlled by a nervous system that causes MUSCLES to contract and relax in a co-ordinated way. The SKELETON provides anchorage for these muscles. To move efficiently through water, land, and air, animals have special adaptations, such as fins, legs, and wings.

Sand is not easy to cross because it shifts under an animal's weight. Side-winder snakes move across soft sand and mud by looping along in S-shaped curves in a movement called side-winding. Instead of slithering across the sand, they throw their bodies through the air in a series of sideways leaps.

Running Gazelle

Ungulates (animals with hooves) are hunted by many predators. Gazelles use their speed and endurance to escape capture. Their lower legs are very long, which lengthens their stride. They also have two toes instead of five, which needs less muscle and so saves energy.

Glider

Several tree-living animals glide from tree to tree using flaps of skin like parachutes. Flying frogs have large, webbed feet that they hold out when they leap so they can fly further without falling to the ground. They can glide up to 15 m (50 ft).

Jet Propulsion

Although fish are strong swimmers, many other marine animals drift along at the mercy of the ocean currents. Jellyfish are able to control their movement to a limited extent. They have a ring of muscle around the edge of their bell-shaped body, which can be contracted and relaxed, like an umbrella opening and closing. This pushes the water backwards, making the jellyfish move in the opposite direction.

Flying

Insects are the smallest animals capable of powered flight. Four-winged insects, such as butterflies, use muscles directly attached to the base of their wings to move the wings up and down. Bees fly by using muscles attached to the top and bottom of their body. When the muscles contract, the wings move upward; when they relax, the wings drop down.

Muscles

Muscles are bundles of fibres that provide the power for animals to move. When a nerve stimulates a muscle into action, the muscles contract (pull back), causing movement. In simple animals, such as snails, muscles contract in waves from one end of the body to the other, pushing the animal along. In vertebrates, such as the horse, muscles work in pairs and pull against bones. The area where different bones meet is called a joint.

High Jump

Fleas need to jump around in order to find an animal from which to suck blood. They can leap an amazing 33 cm (13 in), using muscle energy that is stored in a pad of springy material, called resilin, in their legs. When the leg muscles are triggered to jump, the flea is catapulted into the air.

Skeleton

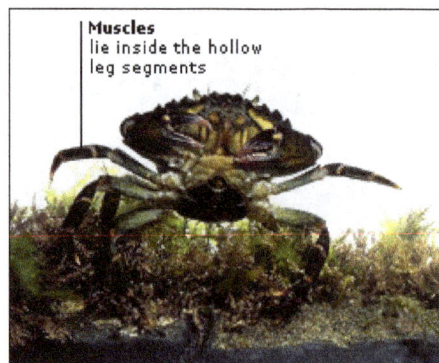

Muscles
lie inside the hollow leg segments

Many animals have a rigid skeleton to support their bodies and some have jointed legs, which allow them to move rapidly. Mammals have the most complex skeletons of all animals. They have backbones made up of many small bones called vertebrae and limbs with several joint types. This complicated skeleton enables them to make lots of different movements.

Animals with exoskeletons, such as crabs, have several pairs of jointed legs. Each pair is made up of a series of hollow sections joined together at joints. Pairs of muscles joined to the inner surface of the joint allow the crab to scuttle sideways quickly.

Streamlined Swimmer

Sharks bodies are specialized for moving fast through water. They have skeletons made from a firm elastic substance called cartilage. Cartilage is lighter than bone, enabling sharks to swim efficiently. Using rhythmic contractions of their body muscle, and with additional push from their tail, they reach speeds of 30–50 kph (19–31 mph).

ANIMAL PHYSIOLOGY

Animal physiology is the study of how animals work and the biological processes essential for animal life, at levels of organization from membranes to the whole animal. It is closely linked with anatomy and with basic physicochemical laws that constrain living as well as nonliving systems. Despite these constraints, there is a diversity of mechanisms and processes by which different animals work. The discipline of animal physiology is underpinned by the concept of homeostasis of the intra- and extracellular environments, neural and endocrine systems for homeostatic regulation, and the various physiological systems including ionic and osmotic balance, excretion, respiration, circulation, metabolism, digestion, and temperature.

ANIMAL CIRCULATORY SYSTEMS

The circulatory system is the primary method used to transport nutrients and gases through the body. Simple diffusion allows some water, nutrient, waste, and gas exchange in animals that are only a few cell layers thick; however, bulk flow is the only method by which the entire body of larger, more complex organisms is accessed.

Circulatory System Architecture

The circulatory system is effectively a network of cylindrical vessels: the arteries, veins, and capillaries that emanate from a pump, the heart. In all vertebrate organisms, as well as some invertebrates, this is a closed-loop system, in which the blood is not free in a cavity. In a closed circulatory system, blood is contained inside blood vessels and circulates unidirectionally from the heart around the systemic circulatory route, then returns to the heart again.

As opposed to a closed system, arthropods– including insects, crustaceans, and most mollusks– have an 'open' circulatory system. In an open circulatory system, the blood is not enclosed in blood vessels but is pumped into an open cavity called a hemocoel and is called hemolymph because the blood mixes with the interstitial fluid. As the heart beats and the animal moves, the hemolymph circulates around the organs within the body cavity and then reenters the hearts through openings called ostia. This movement allows for nutrient exchange, and in some organisms lacking direct gas exchange sites, a basic mechanism to transport gasses beyond the exchange site. Because the gas exchange in many open-circulatory systems tends to be relatively low for metabolically-active organs and tissues, a tradeoff exists between this system and the much more energy-consuming, harder-to-maintain closed system.

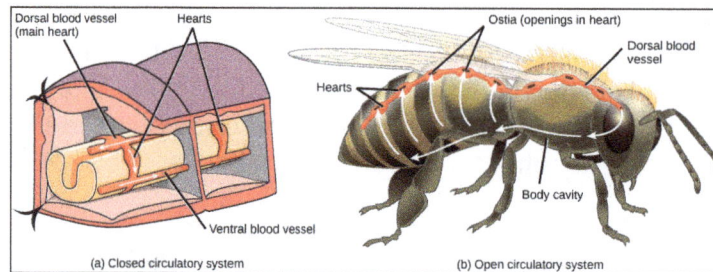

In above figure, (a) closed circulatory systems, the heart pumps blood through vessels that are separate from the interstitial fluid of the body. Most vertebrates and some invertebrates, like this annelid earthworm, have a closed circulatory system. In (b) open circulatory systems, a fluid called hemolymph is pumped through a blood vessel that empties into the body cavity. Hemolymph returns to the blood vessel through openings called ostia. Arthropods like this bee and most mollusks have open circulatory systems.

Circulatory System Variation in Animals

The circulatory system varies from simple systems in invertebrates to more complex systems in vertebrates. The simplest animals, such as the sponges (Porifera) and rotifers (Rotifera), do not need a circulatory system because diffusion allows adequate exchange of water, nutrients, and waste, as well as dissolved gases. Organisms that are more complex but still only have two layers of cells in their body plan, such as jellies (Cnidaria) and comb jellies (Ctenophora) also use diffusion through their epidermis and internally through the gastrovascular compartment. Both their internal and external tissues are bathed in an aqueous environment and exchange fluids by diffusion on both sides. Exchange of fluids is assisted by the pulsing of the jellyfish body.

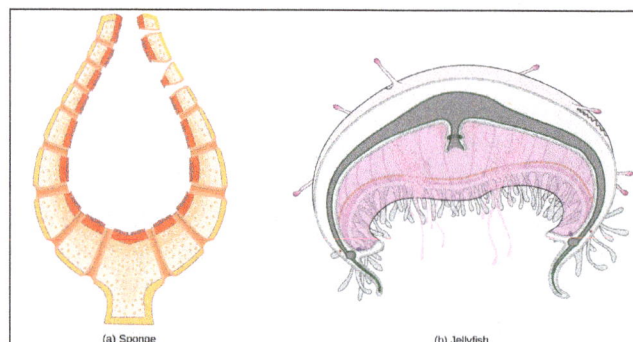

Simple animals consisting of a single cell layer such as the (a) sponge or only a few cell layers such as the (b) jellyfish do not have a circulatory system. Instead, gases, nutrients, and wastes are exchanged by diffusion.

For more complex organisms, diffusion is not efficient for cycling gases, nutrients, and waste effectively through the body; therefore, more complex circulatory systems evolved. In an open system, an elongated beating heart pushes the hemolymph through the body and muscle contractions help to move fluids. The larger more complex crustaceans, including lobsters, have developed arterial-like vessels to push blood through their bodies, and the most active mollusks, such as squids, have evolved a closed circulatory system and are able to move rapidly to catch prey. Closed circulatory systems are a characteristic of vertebrates; however, there are significant differences in the structure of the heart and the circulation of blood between the different vertebrate groups due to adaptation during evolution and associated differences in anatomy. The figure below illustrates the basic circulatory systems of some vertebrates: fish, amphibians, reptiles, and mammals.

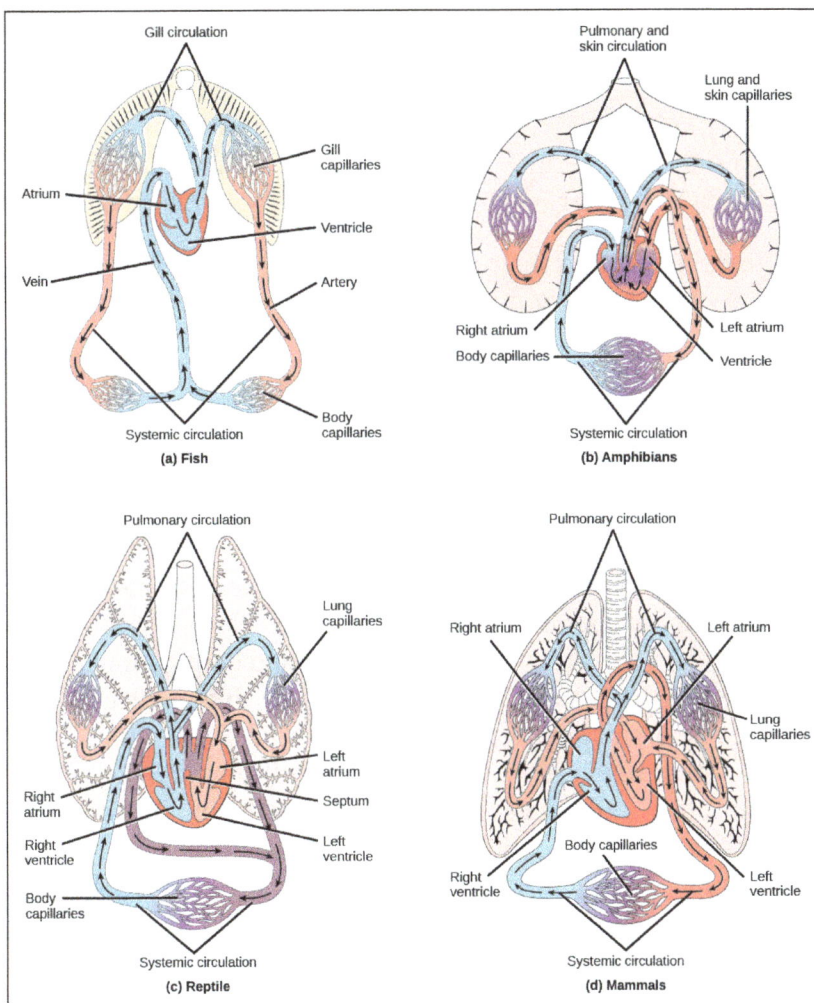

In above figure, (a) Fish have the simplest circulatory systems of the vertebrates: blood flows unidirectionally from the two-chambered heart through the gills and then the rest of the body. (b) Amphibians have two circulatory routes: one for oxygenation of the blood through the lungs and skin, and the other to take oxygen to the rest of the body. The blood is pumped from a three-chambered heart with two atria and a single ventricle. (c) Reptiles also have two circulatory routes; however, blood is only oxygenated through the lungs. The heart is three chambered, but the ventricles are partially separated so some mixing of oxygenated and deoxygenated blood occurs except in croc- odilians and birds. (d) Mammals and birds have the most efficient heart with four chambers that completely separate the oxygenated and deoxygenated blood; it pumps only oxygenated blood through the body and deoxygenated blood to the lungs.

Fish have a single circuit for blood flow and a two-chambered heart that has only a single atrium and a single ventricle. The atrium collects blood that has returned from the body and the ventricle pumps the blood to the gills where gas exchange occurs and the blood is re-oxygenated; this is called gill circulation. The blood then continues through the rest of the body before arriving back at the atrium; this is called systemic circulation. This unidirectional flow of blood produces a gradient of oxygenated to deoxygenated blood around the fish's systemic circuit. The result is a limit in the amount of oxygen that can reach some of the organs and tissues of the body, reducing the overall metabolic capacity of fish.

In amphibians, reptiles, birds, and mammals, blood flow is directed in two circuits: one through the lungs and back to the heart, which is called pulmonary circulation, and the other throughout the rest of the body and its organs including the brain (systemic circulation). In amphibians, gas exchange also oc- curs through the skin during pulmonary circulation and is referred to as pulmocutaneous circulation.

Amphibians have a three-chambered heart that has two atria and one ventricle rather than the two-chambered heart of fish. The two atria (superior heart chambers) receive blood from the two different circuits (the lungs and the systems), and then there is some mixing of the blood in the heart's ventricle (inferior heart chamber), which reduces the efficiency of oxygenation. The advantage to this arrangement is that high pressure in the vessels pushes blood to the lungs and body. The mixing is mitigated by a ridge within the ventricle that diverts oxygen-rich blood through the systemic circulatory system and deoxygenated blood to the pulmocutaneous circuit. For this reason, amphibians are often described as having double circulation.

Most reptiles also have a three-chambered heart similar to the amphibian heart that directs blood to the pulmonary and systemic circuits. However, the ventricle is divided more effectively by a partial septum, which results in less mixing of oxygenated and deoxygenated blood. Some reptiles (alligators and crocodiles) are the most "primitive" animals to exhibit a four-chambered heart. Crocodilians have a unique circulatory mechanism where the heart shunts blood from the lungs toward the stomach and other organs during long periods of submergence, for instance, while the animal waits for prey or stays underwater waiting for prey to rot. One adaptation includes two main arteries that leave the same part of the heart: one takes blood to the lungs and the other pro- vides an alternate route to the stomach and other parts of the body. Two other adaptations include a hole in the heart between the two ventricles, called the foramen of Panizza, which allows blood to move from one side of the heart to the other, and specialized connective tissue that slows the blood flow to the lungs.

In mammals and birds, the heart is divided completely into four chambers: two atria and two ventricles. Oxygenated blood is fully separated from deoxygenated blood, which improves the ef- ficiency of double circulation and is probably required for supporting the warm-blooded lifestyle of mammals and birds. The four-chambered heart of birds and mammals evolved independently from a three-chambered heart. The independent evolution of the same or a similar biological trait is referred to as convergent evolution.

Function and Composition of Blood

Hemoglobin is responsible for distributing oxygen, and to a lesser extent, carbon dioxide, throughout the circulatory systems of humans, vertebrates, and many invertebrates. The blood is more than the proteins, though. Blood is actually a term used to describe the liquid that moves through the vessels and includes plasma (the liquid portion, which contains water, proteins, salts, lipids, and glucose) and the cells (red and white cells) and cell fragments called platelets. Blood plasma is actually the dominant component of blood and contains the water, proteins, electrolytes, lipids, and glucose. The cells are responsible for carrying the gases (red cells) and immune response (white). The platelets are responsible for blood clotting. Interstitial fluid that surrounds cells is separate from the blood, but in hemolymph, they are combined. In humans, cellular components make up approximately 45 percent of the blood and the liquid plasma 55 percent. Blood is 20 percent of a person's extracellular fluid and eight percent of weight.

The Role of Blood in the Body

Blood, like the human blood illustrated below, is important for regulation of the body's systems and homeostasis. Blood helps maintain homeostasis by stabilizing pH, temperature, osmotic pressure, and by eliminating excess heat. Blood supports growth by distributing nutrients and hormones, and by removing waste. Blood plays a protective role by transporting clotting factors and platelets to prevent blood loss and transporting the disease-fighting agents or white blood cells to sites of infection.

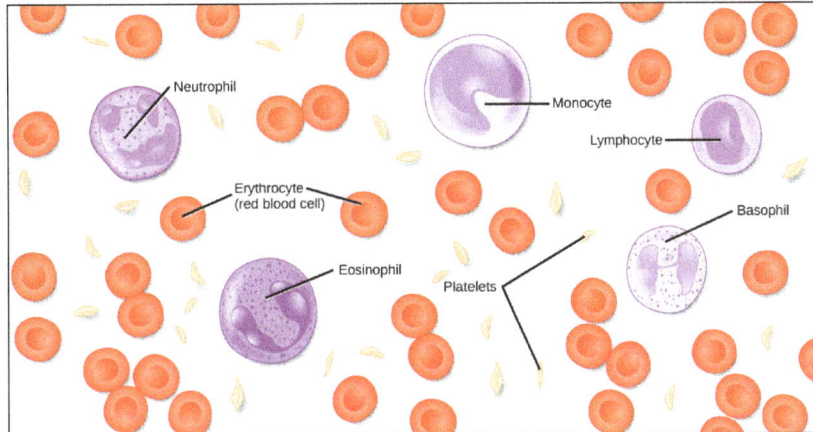

The cells and cellular components of human blood are shown. Red blood cells deliver oxygen to the cells and remove carbon dioxide. White blood cells including neutrophils, monocytes, lymphocytes, eosinophils, and basophils are involved in the immune response.

Red Blood Cells

Red blood cells, or erythrocytes, are specialized cells that circulate through the body delivering oxygen to cells; they are formed from stem cells in the bone marrow. In mammals, red blood cells are small biconcave cells that at maturity do not contain a nucleus or mitochondria and are only 7 µm in size. In birds and non-avian reptiles, a nucleus is still maintained in red blood cells.

The red coloring of blood comes from the iron-containing protein hemoglobin. The principal job of this protein is to carry oxygen, but it also transports carbon dioxide as well. Hemoglobin is packed into red blood cells at a rate of about 250 million molecules of hemoglobin per cell. Each hemoglobin molecule binds four oxygen molecules so that each red blood cell carries one billion molecules of oxygen. There are approximately 25 trillion red blood cells in the five liters of blood in the human body, which could carry up to 25 sextillion ($25^* 10^{21}$) molecules of oxygen in the body at any time. In mammals, the lack of organelles in erythrocytes leaves more room for the hemoglobin molecules, and the lack of mitochondria also prevents use of the oxygen for metabolic respiration. Only mammals have anucleated red blood cells, and some mammals (camels, for instance) even have nucleated red blood cells. The advantage of nucleated red blood cells is that these cells can undergo mitosis. Anucleated red blood cells metabolize anaerobically (without oxygen), making use of a primitive metabolic pathway to produce ATP and increase the efficiency of oxygen transport.

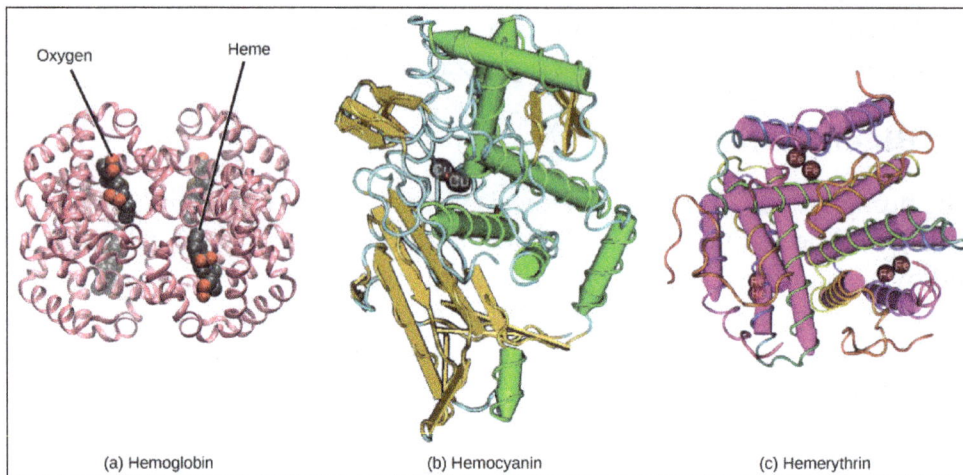

(a) Hemoglobin   (b) Hemocyanin   (c) Hemerythrin

In most vertebrates, (a) hemoglobin delivers oxygen to the body and removes some carbon dioxide. Hemoglobin is composed of four protein subunits, two alpha chains and two beta chains, and a heme group that has iron associated with it. The iron reversibly associates with oxygen, and in so doing is oxidized from $Fe2+$ to $Fe3+$. In most mollusks and some arthropods, (b) hemocyanin delivers oxygen. Unlike hemoglobin, hemolymph is not carried in blood cells, but floats free in the hemolymph. Copper instead of iron binds the oxygen, giving the hemolymph a blue-green color. In annelids, such as the earthworm, and some other invertebrates, (c) hemerythrin carries oxygen. Like hemoglobin, hemerythrin is carried in blood cells and has iron associated with it, but despite its name, hemerythrin does not contain heme.

Not all organisms use hemoglobin as the method of oxygen transport. Invertebrates that utilize hemolymph rather than blood use different pigments to bind to the oxygen. These pigments use copper or iron to the oxygen. Invertebrates have a variety of other respiratory pigments. Hemocyanin, a blue-green, copper-containing protein is found in mollusks, crustaceans, and some of the arthropods. Chlorocruorin, a green-colored, iron-containing pigment is found in four families of polychaete tubeworms. Hemerythrin, a red, iron-containing protein is found in some polychaete worms and annelids. Despite the name, hemerythrin does not contain a heme group and its oxygen-carrying capacity is poor compared to hemoglobin.

The small size and large surface area of red blood cells allows for rapid diffusion of oxygen and carbon dioxide across the plasma membrane. In the lungs, carbon dioxide is released and oxygen is taken in by the blood. In the tissues, oxygen is released from the blood and carbon dioxide is bound for transport back to the lungs. Studies have found that hemoglobin also binds nitrous oxide (NO). NO is a vasodilator that relaxes the blood vessels and capillaries and may help with gas exchange and the passage of red blood cells through narrow vessels. Nitroglycerin, a heart medication for angina and heart attacks, is converted to NO to help relax the blood vessels and increase oxygen flow through the body.

A characteristic of red blood cells is their glycolipid and glycoprotein coating; these are lipids and proteins that have carbohydrate molecules attached. In humans, the surface glycoproteins and glycolipids on red blood cells vary between individuals, producing the different blood types, such as A, B, and O. Red blood cells have an average lifespan of 120 days, at which time they are broken down and recycled in the liver and spleen by phagocytic macrophages, a type of white blood cell.

White Blood Cells

White blood cells, also called leukocytes, make up approximately one percent by volume of the cells in blood. The role of white blood cells is very different than that of red blood cells: they are primarily involved in the immune response to identify and target pathogens, such as invading bacteria, viruses, and other foreign organisms. White blood cells are formed continually; some only live for hours or days, but some live for years.

The morphology of white blood cells differs significantly from red blood cells. They have nuclei and do not contain hemoglobin. The different types of white blood cells are identified by their microscopic appearance after histologic staining, and each has a different specialized function. The two main groups are the granulocytes, which include the neutrophils, eosinophils, and basophils, and the agranulocytes, which include the monocytes and lymphocytes.

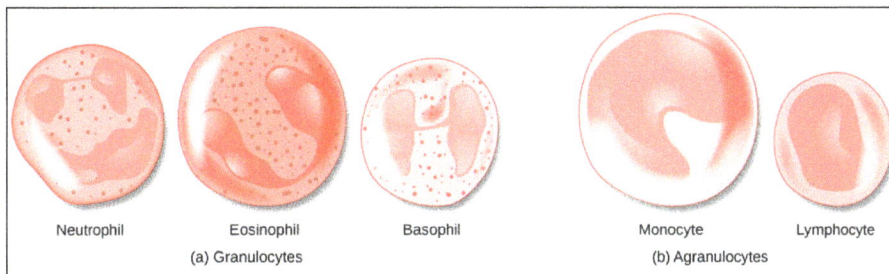

| Neutrophil | Eosinophil | Basophil | Monocyte | Lymphocyte |
| (a) Granulocytes | | | (b) Agranulocytes | |

In above figure, (a) Granulocytes including neutrophils, eosinophils and basophils are characterized by a lobed nucleus and granular inclusions in the cytoplasm. Granulocytes are typically first-responders during injury or infection. (b) Agranulocytes include lymphocytes and monocytes. Lymphocytes, including B and T cells, are responsible for adaptive immune response. Monocytes differentiate into macrophages and dendritic cells, which in turn respond to infection or injury.

Platelets and Coagulation Factors

Blood must clot to heal wounds and prevent excess blood loss. Small cell fragments called platelets (thrombocytes) are attracted to the wound site where they adhere by extending many projections and releasing their contents. These contents activate other platelets and also interact with other coagulation factors, which convert fibrinogen, a water-soluble protein present in blood serum into fibrin (a non-water soluble protein), causing the blood to clot. Many of the clotting factors require vitamin K to work, and vitamin K deficiency can lead to problems with blood clotting. Many platelets converge and stick together at the wound site forming a platelet plug (also called a fibrin clot). The plug or clot lasts for a number of days and stops the loss of blood. Platelets are formed from the disintegration of larger cells called megakaryocytes. For each megakaryocyte, 2000-3000 platelets are formed with 150,000 to 400,000 platelets present in each cubic millimeter of blood. Each platelet is disc shaped and 2-4 ¼m in diameter. They contain many small vesicles but do not contain a nucleus.

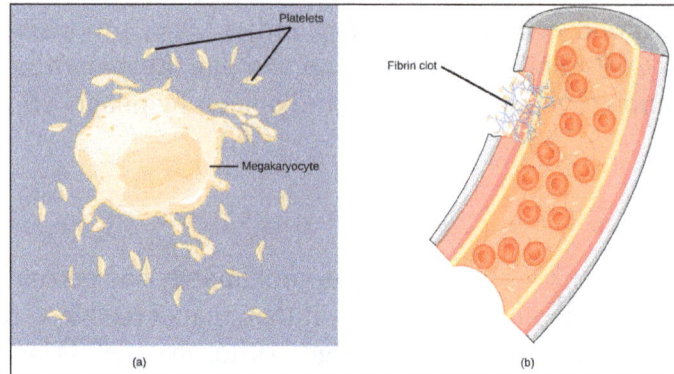

In above figure, (a) Platelets are formed from large cells called megakaryocytes. The megakaryocyte breaks up into thousands of fragments that become platelets. (b) Platelets are required for clotting of the blood. The platelets collect at a wound site in conjunction with other clotting factors, such as fibrinogen, to form a fibrin clot that prevents blood loss and allows the wound to heal.

Functions and Types of Blood Vessels

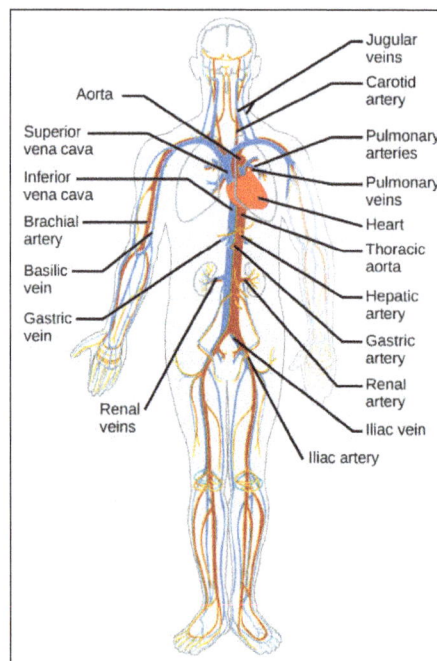

The major human arteries and veins are shown.

The blood from the heart is carried through the body by a complex network of blood vessels. Arteries take blood away from the heart. The main artery is the aorta that branches into major arteries that take blood to different limbs and organs. These major arteries include the carotid artery that takes blood to the brain, the brachial arteries that take blood to the arms, and the thoracic artery that takes blood to the thorax and then into the hepatic, renal, and gastric arteries for the liver, kidney, and stomach, respectively. The iliac artery takes blood to the lower limbs. The major arteries diverge into minor arteries, and then smaller vessels called arterioles, to reach more deeply into the muscles and organs of the body.

Arterioles diverge into capillary beds. Capillary beds contain a large number (10 to 100) of capillaries that branch among the cells and tissues of the body. Capillaries are narrow-diameter tubes that can fit red blood cells through in single file and are the sites for the exchange of nutrients, waste, and oxygen with tissues at the cellular level. Fluid also crosses into the interstitial space from the capillaries. The capillaries converge again into venules that connect to minor veins that finally connect to major veins that take blood high in carbon dioxide back to the heart. Veins are blood vessels that bring blood back to the heart. The major veins drain blood from the same organs and limbs that the major arteries supply. Fluid is also brought back to the heart via the lymphatic system.

The structure of the different types of blood vessels reflects their function or layers. There are three distinct layers, or tunics, that form the walls of blood vessels. The first tunic is a smooth, inner lining of endothelial cells that are in contact with the red blood cells. The endothelial tunic is continuous with the endocardium of the heart. In capillaries, this single layer of cells is the location of diffusion of oxygen and carbon dioxide between the endothelial cells and red blood cells, as well as the exchange site via endocytosis and exocytosis. The movement of materials at the site of capillaries is regulated by vasoconstriction, narrowing of the blood vessels, and vasodilation, widening of the blood vessels; this is important in the overall regulation of blood pressure.

Veins and arteries both have two further tunics that surround the endothelium: the middle tunic is composed of smooth muscle and the outermost layer is connective tissue (collagen and elastic fibers). The elastic connective tissue stretches and supports the blood vessels, and the smooth muscle layer helps regulate blood flow by altering vascular resistance through vasoconstriction and vasodilation. The arteries have thicker smooth muscle and connective tissue than the veins to accommodate the higher pressure and speed of freshly pumped blood. The veins are thinner walled as the pressure and rate of flow are much lower. In addition, veins are structurally different than arteries in that veins have valves to prevent the backflow of blood. Because veins have to work against gravity to get blood back to the heart, contraction of skeletal muscle assists with the flow of blood back to the heart.

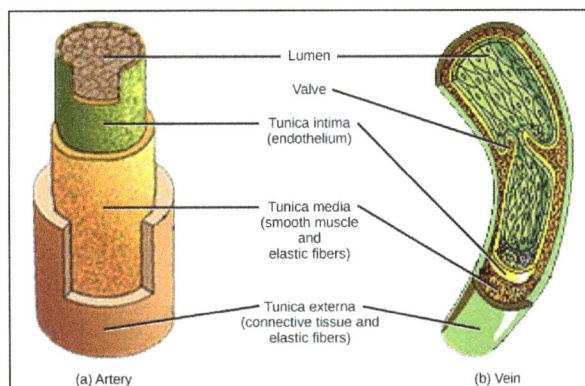

Arteries and veins consist of three layers: an outer tunica externa, a middle tunica media, and an inner tunica intima. Capillaries consist of a single layer of epithelial cells, the tunica intima.

Gas, Nutrient and Fluid Exchange across Blood Vessels

Blood is pushed through the body by the action of the pumping heart. With each rhythmic pump, blood is pushed under high pressure and velocity away from the heart, initially along the main artery, the aorta. In the aorta, the blood travels at 30 cm/sec. As blood moves into the arteries, arterioles, and ultimately to the capillary beds, the rate of movement slows dramatically to about 0.026 cm/sec, one-thousand times slower than the rate of movement in the aorta. While the diameter of each individual arteriole and capillary is far narrower than the diameter of the aorta, and according to the law of continuity, fluid should travel faster through a narrower diameter tube, the rate is actually slower due to the overall diameter of all the combined capillaries being far greater than the diameter of the individual aorta.

The slow rate of travel through the capillary beds, which reach almost every cell in the body, assists with gas and nutrient exchange and also promotes the diffusion of fluid into the interstitial space. After the blood has passed through the capillary beds to the venules, veins, and finally to the main venae cavae, the rate of flow increases again but is still much slower than the initial rate in the aorta. Blood primarily moves in the veins by the rhythmic movement of smooth muscle in the vessel wall and by the action of the skeletal muscle as the body moves. Because most veins must move blood against the pull of gravity, blood is prevented from flowing backward in the veins by one-way valves. Because skeletal muscle contraction aids in venous blood flow, it is important to get up and move frequently after long periods of sitting so that blood will not pool in the extremities.

Blood Pressure and Velocity

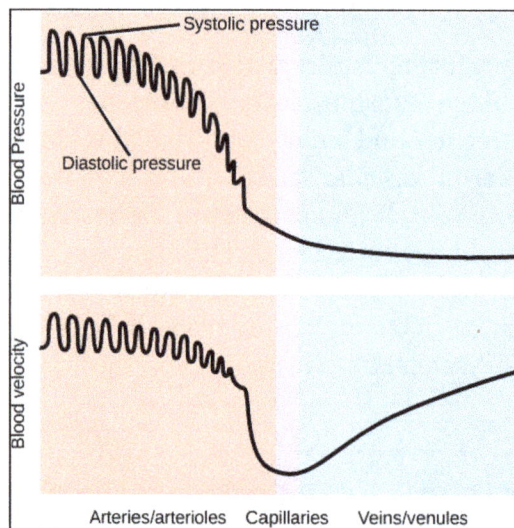

Blood pressure is related to the blood velocity in the arteries and arterioles. In the capillaries and veins, the blood pressure continues to decease but velocity increases.

The pressure of the blood flow in the body is produced by the hydrostatic pressure of the fluid (blood) against the walls of the blood vessels. Fluid will move from areas of high to low hydrostatic pressures. In the arteries, the hydrostatic pressure near the heart is very high and blood flows to the arterioles where the rate of flow is slowed by the narrow openings of the arterioles. During systole, when new blood is entering the arteries, the artery walls stretch to accommodate the increase of pressure of the extra blood; during diastole, the walls return to normal because of their elastic properties. The blood pressure of the systole phase and the diastole phase, graphed below, gives the two pressure readings for blood pressure. For example, 120/80 indicates a reading of 120 mm Hg during the systole and 80 mm Hg during diastole. Throughout the cardiac cycle, the blood continues to empty into the arterioles at a relatively even rate. This resistance to blood flow is called peripheral resistance.

Exchange across Capillaries

Proteins and other large solutes cannot leave the capillaries. The loss of the watery plasma creates a hyperosmotic solution within the capillaries, especially near the venules. This causes about 85% of the plasma that leaves the capillaries to eventually diffuses back into the capillaries near the venules. The remaining 15% of blood plasma drains out from the interstitial fluid into nearby lymphatic vessels. The fluid in the lymph is similar in composition to the interstitial fluid. The lymph fluid passes through lymph nodes before it returns to the heart via the vena cava. Lymph nodes are specialized organs that filter the lymph by percolation through a maze of connective tissue filled with white blood cells. The white blood cells remove infectious agents, such as bacteria and viruses, to "clean" the lymph before it returns to the bloodstream. After it is "cleaned," the lymph returns to the heart by the action of smooth muscle pumping, skeletal muscle action, and one-way valves joining the returning blood near the junction of the venae cavae entering the right atrium of the heart.

ANIMAL DIGESTIVE SYSTEMS

Digestion is the process of breaking down feed into simple substances that can be absorbed by the body. Absorption is the taking of the digested parts of the feed into the bloodstream.

The digestive system consists of the parts of the body involved in chewing and digesting feed. This system also moves the digested feed through the animal's body and absorbs the products of digestion. Different species of animals are better able to digest certain types of feeds than others. This difference occurs because of the various types of digestive systems found in animals. There are four basic types of digestive systems: monogastric, avian, ruminant, and pseudo-ruminant.

Monogastric Digestive System

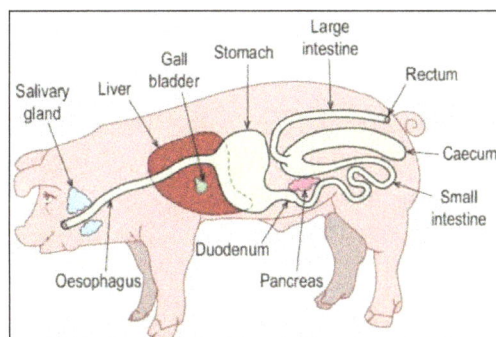

A basic diagram of the digestive system of a hog.

A monogastric digestive system has one simple stomach. The stomach secretes acid, resulting in a low pH of 1.5 to 2.5. The low pH destroys most bacteria and begins to break down the feed materials. Animals with this type of digestive system are better adapted to eat rations high in concentrates. Concentrates are highly digestible feedstuffs that are high in energy and low in fiber. Concentrates are typically 80 to 90 percent digestible. Common concentrates are cereal grains and oil meals. Cereal grains include corn, wheat, barley, and oats. Oil meals include soybean meal, linseed meal, and cottonseed meal. Examples of monogastric animals are hogs, cats, dogs, and humans.

Avian Digestive System

The avian digestive system is found in poultry. This system differs greatly from any other type. Since poultry do not have teeth, there is no chewing. Poultry break their feed into pieces small enough to swallow by pecking with their beaks or scratching with their feet. Feed enters the mouth, travels to the esophagus, and empties directly into the crop. The crop is where the food is stored and soaked. Food then moves from the crop to the proventriculus. The proventriculus is the stomach in a bird, where gastric enzymes and hydrochloric acid are secreted. From the proventriculus, the food makes its way to the gizzard. The gizzard is a very muscular organ, which normally contains grit or stones that function like teeth to grind the food. The food then moves from the gizzard to the small intestine and then to the large intestine. The nondigestible food components then travel into the cloaca. Urine is also emptied into the cloaca. The material is then passed out of the body through the vent. Digestion in the avian system is very rapid.

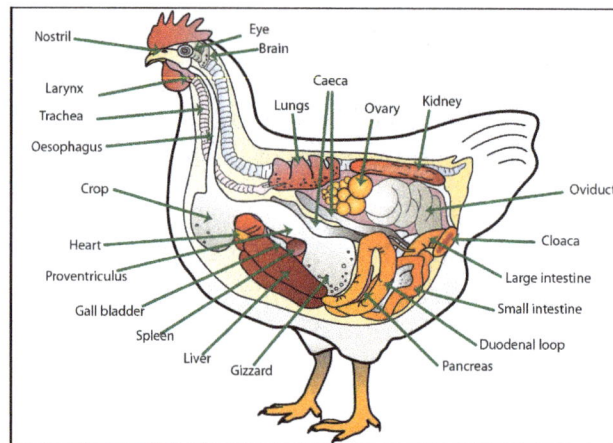

A basic diagram of the digestive system of a chicken.

Ruminant Digestive System

The ruminant digestive system has a large stomach divided into four compartments—the rumen, the reticulum, the omasum, and the abomasum. The ruminant digestive system is found in cattle, sheep, goats, and deer. Ruminant animals eat feed rations that are high in roughages and low in concentrates. Roughages are feedstuffs that are high in fiber, low in energy, and typically only 50 to 65 percent digestible. Roughages include hay, straw, grazed forages, and silage. Ruminants are different from monogastric animals in that they swallow their food in large quantities with little chewing. Later they will ruminate, or belch up the feed, chew, and swallow it again. The regurgitated feed is called a cud. A cud is a ball-like mass of feed brought up from the stomach to be rechewed. On average, cattle chew their cuds about six to eight times per day.

Rumen

The first and largest section of the stomach is the rumen. In the rumen, solid feed is mixed and partially broken down. The rumen contains millions of bacteria and other microbes that promote fermentation, which breaks down roughages. The rumen also contains microorganisms that synthesize amino acids and B-complex vitamins. Amino acids are the building blocks of proteins and are essential for the growth and maintenance of cells.

Reticulum

The reticulum is the second segment of the stomach. The reticulum is a small pouch on the side of the rumen that traps foreign materials, such as wire, nails, and so forth. Since ruminants do not chew their food before swallowing, they will occasionally swallow foreign objects.

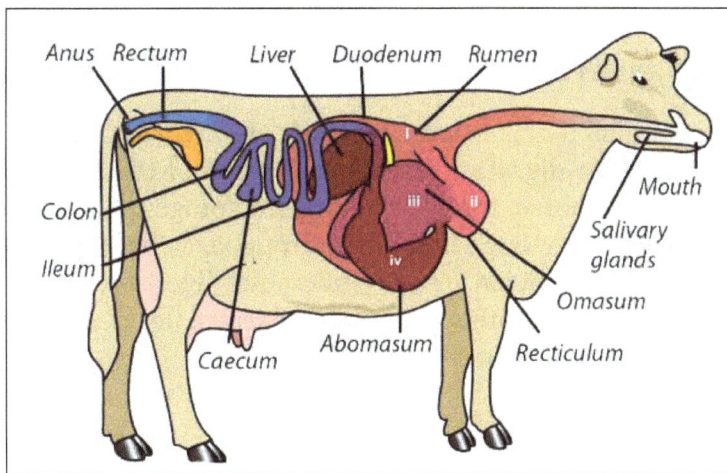

A basic diagram of the digestive system of a cow.

Omasum

The omasum is the third compartment of the stomach. The omasum produces a grinding action on the feed and removes some of the water from the feed. Hydrochloric acid and digestive enzymes are mixed with feed in the omasum.

Abomasum

The abomasum is the fourth compartment of the stomach. The abomasum is also referred to as the true stomach because it is similar to the stomach in monogastric animals.

Rumen Microorganisms

Ruminants rely on microorganisms for the digestion of roughages. The rumen microorganisms are very diverse and consist of bacteria, protozoa, and fungi.

Bacteria are the most numerous rumen microorganisms, at approximately 1 billion bacteria per milliliter of rumen fluid. Bacteria are responsible for most feed digestion in the rumen. They break down cellulose to form volatile fatty acids (VFAs). The VFAs provide the ruminant with 60 to 80 percent of its energy needs. Protozoa are typically responsible for about 25 percent of the fiber digestion in the rumen, even though a ruminant can survive without any protozoa in the rumen. Fungi contribute up to 8 percent of the total rumen microorganisms. Fungi are responsible for the digestion of cellulose and lignin in more resistant forages, such as barley straw.

Feed conversion and rate of gain in a ruminant are strongly affected by the type and number of microorganisms in the rumen. The rumen must contain the appropriate proportions of certain types of microorganisms to maximize productivity. For example, it is believed that protozoa can have a negative impact on protein utilization. The number of protozoa in the rumen is inversely proportionate to the number of bacteria. Therefore, if a ruminant is fed in a manner that is most conducive to bacteria in the rumen, protein utilization will be maximized by eliminating or reducing the number of protozoa in the rumen. To illustrate, feeding yeast culture to cattle could help to ensure a healthy population of rumen bacteria.

Pseudo-ruminant Digestive System

A pseudo-ruminant is an animal that eats large amounts of roughage but does not have a stomach with several compartments. The digestive system does some of the same functions as those of ruminants. For example, in the horse, the cecum ferments forages. An animal with a pseudo-ruminant digestive system can utilize large amounts of roughages because of the greatly enlarged cecum and large intestine, which provide areas for microbial digestion of fiber. Pseudo-ruminants often eat forages as well as grains and other concentrated feeds. Besides horses, examples of pseudo-ruminants are rabbits, guinea pigs, and hamsters.

ANIMAL RESPIRATORY SYSTEM

Every cell in an animal requires oxygen to perform cellular respiration. Cellular respiration is the process by which animals take in oxygen and exchange it for carbon dioxide and water as waste products. Animals have specialized systems that help them do this successfully and efficiently. Even a fish will drown if it can't breathe underwater.

Gas Exchange

The actual exchanging of gases is dependent upon important structures such as lungs or gills, and the principle of diffusion. Diffusion is the process where molecules or particles move from an area where they are very concentrated into an area where they are less concentrated.

Below is an illustration showing the process of diffusion. A container is separated with a semipermeable membrane, dividing an area with a high concentration of a molecule (red dots) from an area with lower concentration. The membrane allows the molecules to move from one side to the other. Over time, the molecules will move from an area of higher concentration to an area of lower concentration. The molecules will continue to migrate across the membrane until there is an equal amount on both sides. This is called equilibrium, or when both sides of a membrane have equal concentrations.

This is relevant during respiration because oxygen and carbon dioxide are often highly concentrated on opposite sides of a cell membrane. Diffusion allows gas exchange to occur.

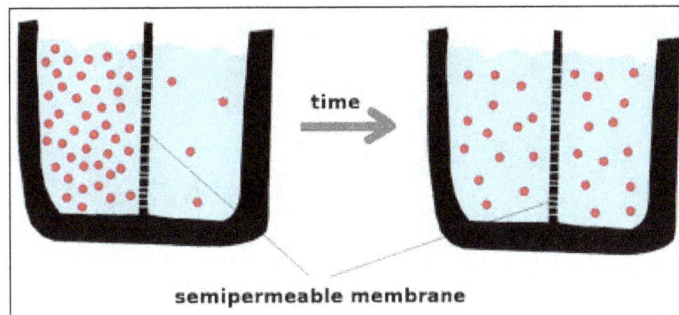

semipermeable membrane

A diagram demonstrating diffusion.

In animals with a closed circulatory system (such as birds, mammals, reptiles, and some amphibians), gas exchange takes place across the capillaries. Remember that the capillaries are the smallest blood vessel and can be found near every cell in the body. With trillions of cells, that's a lot of capillaries.

Insects

The respiratory system in insects consists of a network of tubes, called tracheae, which directly ventilate the tissues. Actively moving air to the site of gas exchange is called ventilation. The tubes divide and branch out into smaller and smaller tubes extending into all parts of the insect, similar to the way arteries branch out into tiny capillaries in a closed circulatory system.

Insects have openings scattered throughout its body called spiracles. Spiracles are openings to the tracheae. In small insects, gas exchange occurs by diffusion only. Larger insects will actively breathe to pump air into the tubes.

Aquatic insects must seal their spiracles when they are under water to prevent flooding their tubes. Amazingly, some aquatic insects even have specialized spiracles that can puncture underwater plants and access those plants' oxygen storage centers. Think of it like an underwater vampire bug that sucks oxygen.

Mammals

The chief organ in mammalian respiration is the lungs. The lungs are actively ventilated via a suction-pump mechanism of inhalation and exhalation. Breathing is dependent upon the rib muscles and the diaphragm, a structure shaped like a dome-shaped floor just beneath the lungs.

Inhalation happens when the rib cage opens up and the diaphragm flattens and moves downward. The lungs expand into the larger space, causing the air pressure inside to decrease. The drop in air pressure inside the lung makes the outside air rush in.

Exhalation is the opposite process. The diaphragm and the rib muscles relax to their neutral state, causing the lungs to contract. The squashing of the lungs increases their air pressure and forces the air to flow out.

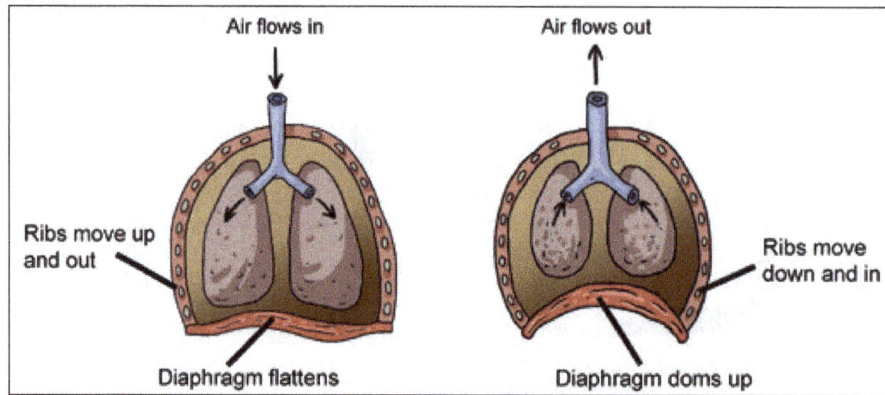

A diagram of ventilation in most mammals. The left image shows
inhalation with a flattened diaphragm. The right side shows the
dome-shaped diaphragm forcing the air out during exhalation.

Most mammals are nose breathers. Inhaling through the nose warms and moistens the air. The air is filtered by cilia and mucus membranes, which trap dust and pathogens. Air then reaches the epiglottis, the tiny leaf-shaped flap at the back of the throat. The epiglottis regulates air going into the windpipe and closes upon swallowing to prevent food from being inhaled. It's the gatekeeper to the lungs.

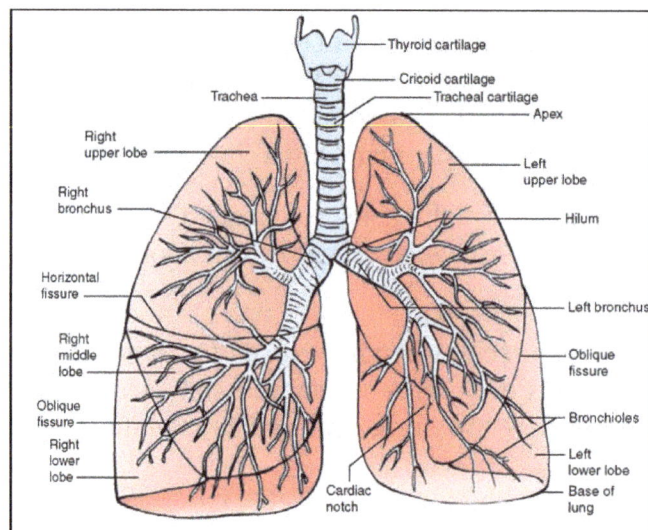

Diagram of structures of the lungs.

The trachea is a long structure of soft tissue surrounded by c-shaped rings of cartilage. In humans, the trachea splits into two bronchi branches that lead to each lung. Each bronchi divides into increasingly smaller branches, until they form a massive tree of tubes. The smallest branches are called the bronchioles, and each bronchiole ends with a tiny air sac (no larger than a grain of sand) called an alveolus.

The tiny alveoli (plural of alveolus) are crucial because they increase the surface area used for gas exchange. If the lungs were just empty sacs, then only area available for gas exchange would be the walls of the lungs. In humans, that comes out to an area of approximately 0.01 m2. The alveoli, though, provide a whopping 75m2 of surface area where oxygen absorption can take place. That's the size of half a volleyball court and it's all inside of you.

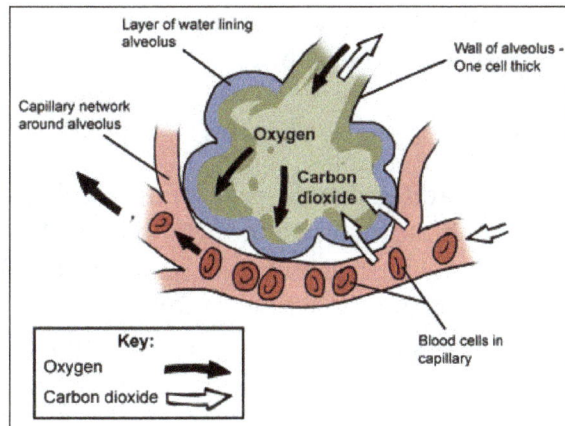

Diagram of an alveolus near a capillary and the gas exchange process in the lungs.

Gas exchange takes place in the capillaries, so the alveoli have a close working relationship with the network of capillaries. This brings the blood-carrying waste products close enough to the fresh air for diffusion to take place. The waste is removed and the oxygen is taken up by the blood. The hemoglobin in blood attaches oxygen molecules, kind of like a bus carrying passengers. Each hemoglobin protein can carry four passengers of oxygen at one time. Oxygen is delivered to the cells and carbon dioxide is removed. Water vapor and carbon dioxide are exhaled, and the process begins again with inhalation.

Just as the heart beats on its own, breathing is done without conscious effort. There are sections of the brain, called the medulla and pons that regulate respiration. They control the rate of respiration by monitoring carbon dioxide levels in the blood. In times of excitement or during exercise, the cells require more oxygen than normal and respiration speeds up.

Tidal volume is the amount of air breathed in or out during a respiratory cycle. The tidal volume and respiratory frequency, or the number of inhalations over a period of time (such as breaths per minute), vary amongst species, and it's affected by age, pregnancy, exercise, excitement, temperature, and body size. For instance, horses have an average respiration of 12 times per minute, but pigs breathe an average of 40 times per minute.

Horses are obligate nasal breathers, meaning that they can only breathe through their noses and are unable to breathe through their mouths. Humans and many other mammals can breathe through either their mouths or their noses. It's thought that this modification allows horses to graze with their heads down while separate nasal passages breathe in air and sniff for potential predators.

Marine mammals breathe oxygen with lungs just like their terrestrial brethren, but with a few differences. To prevent water from getting into their airway, they have adapted muscles or cartilaginous flaps to seal their tracheas when under the water. We wish our cartilaginous flaps did that, but our ancestors skipped out on a couple million years of pruning up in the ocean, so we're out of luck. Marine mammals also exchange up to 90% of their gases in a single breath, which helps them gather as much oxygen and expel as much waste as possible. A sperm whale can last for 138 minutes on a single breath.

It can be dangerous for diving mammals to have air in their lungs when they dive to great depths. The water pressure would exert too much force on the air in their lungs, causing them to burst. For this reason, many marine mammals will prepare for a deep dive by taking a breath, exchanging gases in the blood, and exhaling to empty their lungs.

Reptiles and Amphibians

Reptiles and amphibians have lungs and exchange gases in the capillaries like mammals, but there are some differences in how they ventilate their respiratory systems. Reptiles don't typically breathe the same way as mammals, since many reptiles lack a diaphragm. Reptiles use their axial muscles, the ones attached to their ribs, to expand their ribcage for breathing. During periods of intense activity, reptiles might be forced to hold their breath, as they use those muscles for running away.

Some reptiles get around this by buccal pumping while they run. Buccal pumping is when an animal uses the muscles of the mouth and throat to pull air into the lungs. Muscles pull air through the mouth or nose into a buccal cavity. Throat muscles then pump and move the floor of the mouth up in a way that's visible from the outside. This forces air out of the mouth and into the lungs. This is what amphibians do, by puffing up their chinny-chin-chins to get the air in.

Apart from their capillaries, amphibians perform gas exchange directly through their skin. This works for them because their skin has lots of blood vessels very close to the permeable skin surface. Diffusion can take place right through the skin. In fact, some salamanders have no lungs at all, and they get all of their oxygen through their skin.

Birds

The respiratory system of birds is similar to that of mammals. Air is pulled in using a suction-type pull. Gases are exchanged in the capillaries. The major difference is the route of airflow through the body. Birds have air sacs that collect air. They then force the air through their lungs like bellows stoking a fire.

When a bird inhales, air is brought into the posterior air sacs, which expand. Then the bird exhales and the air is forced from the posterior air sacs into the lungs, where gas exchange occurs. The bird inhales a second time, moving the air from the lungs to the anterior air sac. A second exhalation pushes the air out of the body.

This progression of air through the bird means that the lungs are compressed during inhalation and expand during exhalation. It also takes two full inhalations and exhalations to move one gulp of air through the bird. That's a lot of gulps.

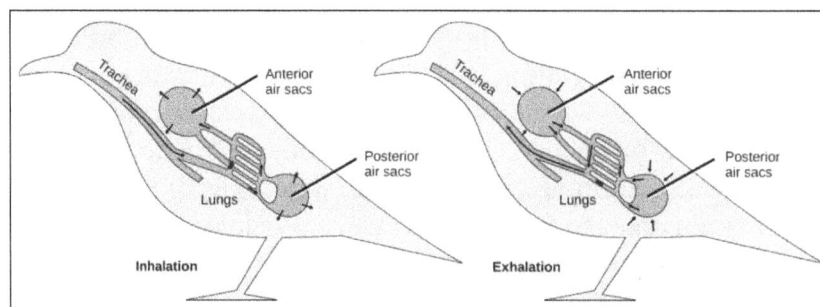

Diagram of the ventilation process in avian respiration. It shows the air going into an air sac before it reaches the lungs and again after it passes through the lungs.

The unidirectional flow allows all the air flowing through the lungs to be fresh air with maximal oxygen to be collected. In humans, this is not the case since there's only one pathway to the lungs and it's used for both entry and exit. During flight, air sacs and lungs are continuously filled with oxygen rich air. This provides maximal oxygen to be absorbed into the blood stream, which is necessary for the high metabolism needed for flight.

Aquatic Respiration

In fish, respiration takes place in the gills. Gills collect dissolved oxygen from the water and release carbon dioxide. Gills are much more complex than just a slit in the cheeks of a fish.

Gills are comprised of gill arches with hundreds of gill filaments extending from them. Each filament is lined with rows of lamellae, and the gas exchange takes place as water flows through them. The frills and flaps increase the surface area to allow more gas exchange to take place, just as the alveoli do in the lungs.

Fish utilize a countercurrent exchange pathway (except for cartilaginous fish), which means that their arteries are arranged so that blood flows in the opposite direction of water movement against the gills. By having their respiration pathway in this orientation, maximum gas exchange can take place.

If the blood and the water were moving in the same direction, the blood would always be next to the same bit of water which would soon be depleted of oxygen. By setting up a countercurrent pathway, the blood is always passing water that is still oxygenated. This allows the blood to gather as much oxygen as possible.

Since water must be flowing over the gills to provide a continual source of oxygen, fish have developed several ways to keep them ventilated. Some fish swim with their mouths open almost all of the time. Other fish have a special flap called an operculum, which is used to force water across the gills.

Like all good rules, there's an exception. While all fish have gills, one fish also has lungs. The lungfish can survive when its water habitat dries up from seasonal drought. What an aptly named fish. There's also certain land crabs that have both lungs and gills, and can breathe both under the sea and on land.

The lungfish is a unique animal which has gills and lungs.

ANIMAL EXCRETORY SYSTEM

Cells produce water and carbon dioxide as by-products of metabolic breakdown of sugars, fats, and proteins. Chemical groups such as nitrogen, sulfur, and phosphorous must be stripped, from the large molecules to which they were formerly attached, as part of preparing them for energy conversion. The continuous production of metabolic wastes establishes a steep concentration gradient across the plasma membrane, causing wastes to diffuse out of cells and into the extracellular fluid.

Single-celled organisms have most of their wastes diffuse out into the outside environment. Multicellular organisms, and animals in particular, must have a specialized organ system to concentrate and remove wastes from the interstitial fluid into the blood capillaries and eventually deposit that material at a collection point for removal entirely from the body.

Regulation of Extracellular Fluids

Excretory systems regulate the chemical composition of body fluids by removing metabolic wastes and retaining the proper amounts of water, salts, and nutrients. Components of this system in vertebrates include the kidneys, liver, lungs, and skin.

Not all animals use the same routes or excrete their wastes the same way humans do. Excretion applies to metabolic waste products that cross a plasma membrane. Elimination is the removal of feces.

Nitrogen Wastes

Nitrogen wastes are a by product of protein metabolism. Amino groups are removed from amino acids prior to energy conversion. The NH_2 (amino group) combines with a hydrogen ion (proton) to form ammonia (NH_3).

Ammonia is very toxic and usually is excreted directly by marine animals. Terrestrial animals usually need to conserve water. Ammonia is converted to urea, a compound the body can tolerate at higher concentrations than ammonia. Birds and insects secrete uric acid that they make through large energy expenditure but little water loss. Amphibians and mammals secrete urea that they form in their liver. Amino groups are turned into ammonia, which in turn is converted to urea, dumped into the blood and concentrated by the kidneys.

Water and Salt Balance

The excretory system is responsible for regulating water balance in various body fluids. Osmoregulation refers to the state aquatic animals are in: they are surrounded by freshwater and must constantly deal with the influx of water. Animals, such as crabs, have an internal salt concentration very similar to that of the surrounding ocean. Such animals are known as osmoconformers, as there is little water transport between the inside of the animal and the isotonic outside environment.

Marine vertebrates, however, have internal concentrations of salt that are about one-third of the surrounding seawater. They are said to be osmoregulators. Osmoregulators face two problems: prevention of water loss from the body and prevention of salts diffusing into the body. Fish deal with this by passing water out of their tissues through their gills by osmosis and salt through their gills by active transport. Cartilaginous fish have a greater salt concentration than seawater, causing water to move into the shark by osmosis; this water is used for excretion. Freshwater fish must prevent water gain and salt loss. They do not drink water, and have their skin covered by a thin mucus. Water enters and leaves through the gills and the fish excretory system produces large amounts of dilute urine.

Terrestrial animals use a variety of methods to reduce water loss: living in moist environments, developing impermeable body coverings, production of more concentrated urine. Water loss can be considerable: a person in a 100 degree F temperature loses 1 liter of water per hour.

Excretory System Functions

- Collect water and filter body fluids.

- Remove and concentrate waste products from body fluids and return other substances to body fluids as necessary for homeostasis.

- Eliminate excretory products from the body.

Invertebrate Excretory Organs

Many invertebrates such as flatworms use a nephridium as their excretory organ. At the end of each blind tubule of the nephridium is a ciliated flame cell. As fluid passes down the tubule, solutes are reabsorbed and returned to the body fluids.

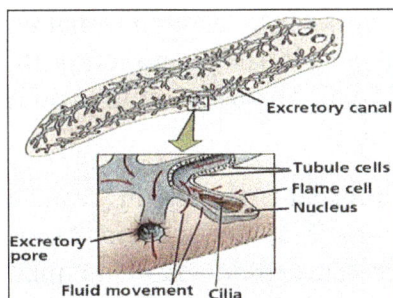

Excretory system of a flatworm.

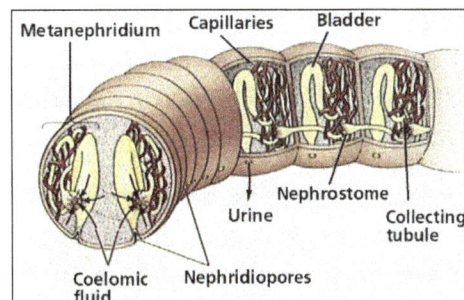

Excretory system of an earthworm.

Body fluids are drawn into the Malphigian tubules by osmosis due to large concentrations of potassium inside the tubule. Body fluids pass back into the body, nitrogenous wastes empty into the insect's gut. Water is reabsorbed and waste is expelled from the insect.

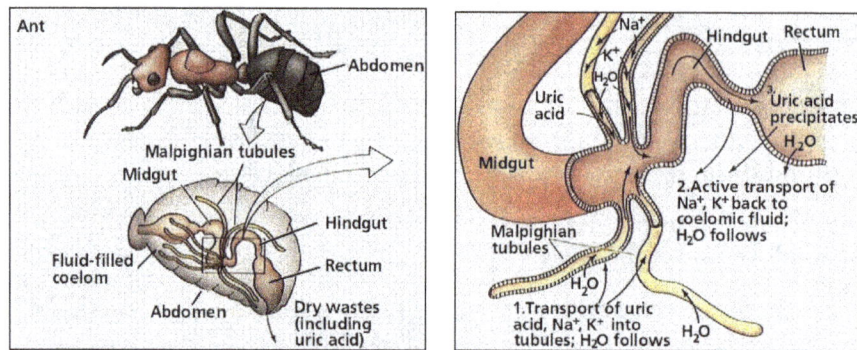

Excretory system of an ant.

Vertebrates have Paired Kidneys

ALL vertebrates have paired kidneys. Excretion is not the primary function of kidneys. Kidneys regulate body fluid levels as a primary duty, and remove wastes as a secondary one.

HOMEOSTASIS

Environmental factors are the characteristics of the external environment that have a direct impact on our organism. Sometimes, the alteration of some of these parameters (such as temperature or humidity) can negatively affect bodies, damaging and disrupting the execution of physiological functions. It's important for animals to be able to have control over these factors, in order to maintain a certain degree of constance and stability in their organism.

Some regular activities, such as drinking or transpiring, help achieve this stability, but there are a series of extremely complex mechanisms involved in the regulation of the body. Keep on reading to learn about the importance of homeostatic mechanisms, which are in charge of maintaining a state of steady internal conditions.

The Importance of Maintaining a Constant Internal Environment

The cells that form the organs and tissues of animals are immersed in a liquid medium, a fluid compartment that Claude Bernard, father of modern physiology, called 'internal environment'. The internal environment refers, mainly, to extracellular fluid (ECF), a section that separates blood from cells, which is, in turn, composed by interstitial fluid, plasma liquid and lymph, fluids that are crucial in the performance of physiological functions.

By researching mammals, Bernard discovered that this internal environment stayed considerably steady, even when there were fluctuations in external parameters; the variation of several environmental factors, such as temperature or environmental pressure, didn't cause an imbalance in the composition and properties of the internal environment, which remained stable.

Animals that maintain their internal environment
steady are able to exploit a wider variety of habitats.

The discovery of the constance of the internal environment was extremely significant, as it allowed researchers to reach the conclusion that animals that were able to regulate their internal environment were also able to exploit a wider variety of potential habitats. This revelation allowed Bernard to formulate one of his most famous statements: "the constance of the internal environment is the condition for a free and independent life", meaning that those beings that are able to maintain the constance of the internal environment can be considered an organism that is independent of the environment. In any case, and in order to make this happen, a mechanism known as homeostasis is involved.

Homeostasis

In 1932, Cannon defined homeostasis as the series of physiological processes that are involved in the regulation and maintenance of the state of an organism in the face of any disturbance. It's important to note that the main destabilizing factors of the internal environment are environmental parameters and the cellular metabolism itself.

The homeostatic processes involve a series of internal sensors (sensory receptors) that can detect any kind of deviation from an optimal physiological state, and, at the same time, to initiate the appropriate actions to correct these alterations.

This optimal state can be maintained by set point, that is, by an appropriate reference value for each species: when a disturbance (vibrations, radiation...) is perceived by the sensory receptors, the organism checks that reference value and sets the appropriate homeostatic mechanisms, which act in consequence to maintain that value. Homeostasis includes both physiological and ethological mechanisms: sweating, panting (physiological thermoregulation), occultation, fur (ethological responses to cold), etc. In short, homeostatic mechanisms are essential for animals, as they regulate and maintain the organism in optimum conditions, even when they face adversity. For example, it has been proven that, in some rodents, the blood sugar levels remain constant, even when they have no access to food.

The Methods of Homeostatic Regulation

Two different homeostatic mechanisms are in charge of maintaining the stability of the internal environment.

Reactive Homeostasis

Babbling is a physiological mechanism.

Reactive homeostasis is a direct response to the changes that take place in the internal environment (a variation in pH, for example); that is, it occurs when an internal parameter of the organism is subject to a variation that must be corrected. An example of reactive homeostasis is the moment when an animal drinks as a response to a dehydration caused by excessive panting or heavy sweating.

Predictive Homeostasis

The internal oscillating mechanisms act as true chronometers, which can prepare a physiological response to external environmental changes in advance. This early preparation is known as 'predictive homeostasis', a term proposed by Martin Moore-Ede.

Predictive homeostasis is a response to changes in the external environment. It's anticipatory, meaning that it allows to predict the appearance of an environmental stimulus, and to anticipate a proper response to any disturbance that will divert the reference value or set point. This model of homeostasis also affects the circadian system, which, aware of the disturbance, allows the deviation of the reference value, so the organism must regulate from this new, modified set point (the adaptive response acts in reference to the new set point).

Some types of macaws provide an interesting example of predictive homeostasis: this group of birds often consumes a clay mineral called 'kaolin', which acts as a natural drug that prevents potential intoxications by the ingestion of seeds. Another example would be the reduction of food intake by animals that are dehydrated, in order to avoid the losing water through excretion.

The red-and-green macaw (Ara chloropterus) is able to anticipate the changes of the external environment.

Types of Organisms and their Regulatory Mechanisms

There are different types of organisms, depending on the regulatory mechanisms they use. In general, we can say that, as we ascend in the evolutionary scale, the ability to maintain the stability of the internal environment will be more and more effective, making the process of homeostasis increasingly sophisticated.

Conformer Organisms

Conformer species are influenced by external factors, so the organism gradually adapts its internal parameters to the environmental parameters, thanks to the flexibility of its enzymes. Conformers have an advantage: they don't have to invest so much energy in keeping internal characteristics stable. However, the possibilities of free life are limited, as internal cells are subject to the alterations of external conditions.

Salmons are conformers, as their body temperature
adapts to the waters they inhabit.

These organisms can only perform their functions satisfactorily in a narrow margin of parameters, while, outside that range, they simply try to survive. In general, conformers tolerate wide variations in the parameters of their internal environment.

Regulator Organisms

Regulator organisms maintain the conditions of their internal environment stable, within narrow limits, in the face of the variation of the conditions of the external environment. Unlike conformers, the cells of regulators work independently from the external environment, tolerating extensive changes in its characteristics. The mechanisms that make this possible consume a lot of energy. Mammals, for example, are regulatory organisms.

References

- Animal-anatomy, science, encyclopedia, dk: factmonster.com, Retrieved 9 August, 2019

- Movement, science, encyclopedia, dk: factmonster.com, Retrieved 10 January, 2019

- Animal-circulatory-systems, nutrition-transport-and-homeostasis: biology.gatech.edu, Retrieved 11 February, 2019

- Animal-respiration, animal-movement: shmoop.com, Retrieved 12 March, 2019

- What-is-homeostasis: zooportraits.com, Retrieved 13 April, 2019

6

Animal Behavior

The scientific study of animal behavior takes place under the domain of ethology. It views their behavior as an evolutionary adaptive trait. This chapter has been carefully written to provide an easy understanding of the varied facets of animal behavior such as their learned behavior, adjunctive behavior, social behavior, territorial behavior and sexual behavior.

Animal Behavior refers to everything animals do, including movement and other activities and underlying mental processes. Human fascination with animal Behavior probably extends back millions of years, perhaps even to times before the ancestors of the species became human in the modern sense. Initially, animals were probably observed for practical reasons because early human survival depended on knowledge of animal Behavior. Whether hunting wild game, keeping domesticated animals, or escaping an attacking predator, success required intimate knowledge of an animal's habits. Even today, information about animal Behavior is of considerable importance. For example, in Britain, studies on the social organization and the ranging patterns of badgers (Meles meles) have helped reduce the spread of tuberculosis among cattle, and studies of sociality in foxes (Vulpes vulpes) assist in the development of models that predict how quickly rabies would spread should it ever cross the English Channel. Likewise in Sweden, where collisions involving moose (Alces alces) are among the most common traffic accidents in rural areas, research on moose Behavior has yielded ways of keeping them off roads and verges. In addition, investigations of the foraging of insect pollinators, such as honeybees, have led to impressive increases in agricultural crop yields throughout the world.

Even if there were no practical benefits to be gained from learning about animal Behavior, the subject would still merit exploration. Humans (Homo sapiens) are animals themselves, and most humans are deeply interested in the lives and minds of their fellow humans, their pets, and other creatures. British ethologist Jane Goodall and American field biologist George Schaller, as well as British broadcaster David Attenborough and Australian wildlife conservationist Steve Irwin, have brought the wonders of animal Behavior to the attention and appreciation of the general public. Books, television programs, and movies on the subject of animal Behavior abound.

Darwin's Influence

The origins of the scientific study of animal Behavior lie in the works of various European thinkers of the 17th to 19th centuries, such as British naturalists John Ray and Charles Darwin and French naturalist Charles LeRoy. These individuals appreciated the complexity and apparent purposefulness of the actions of animals, and they knew that understanding Behavior demands long-term observations of animals in their natural settings. At first, the principal attraction of natural history studies was to confirm the ingenuity of God. The publication of Darwin's On the Origin of Species in 1859 changed this attitude. In his chapter on instinct, Darwin was concerned with whether behavioral traits, like anatomical ones, can evolve as a result of natural selection. Since then, biologists have recognized that the Behaviors of animals, like their anatomical structures, are adaptations that exist because they have, over evolutionary time (that is, throughout the formation of new species and the evolution of their special characteristics), helped their bearers to survive and reproduce.

Charles Darwin.

Furthermore, humans have long appreciated how beautifully and intricately the Behaviors of animals are adapted to their surroundings. For example, young birds that possess camouflaged colour patterns for protection against predators will freeze when the parent spots a predator and calls the alarm. Darwin's achievement was to explain how such wondrously adapted creatures could arise from a process other than special creation. He showed that adaptation is an inexorable result of four basic characteristics of living organisms:

- There is variation among individuals of the same species. Even closely related individuals, such as parent and offspring or sibling and sibling, differ considerably. Familiar human examples include differences in facial features, hair and eye colour, height, and weight.

- Many of these variations are inheritable—that is, offspring resemble their parents in many traits as a result of the genes they share.

- There are differences in numbers of surviving offspring among parents in every species. For example, one female snapping turtle (family Chelydridae) may lay 24 eggs; however, only 5 may survive to adulthood. In contrast, another female may lay only 18 eggs, with 1 of her offspring surviving to adulthood.

- The individuals that are best equipped to survive and reproduce perpetuate the highest frequency of genes to descendant populations. This is the principle known colloquially as "survival of the fittest," where fitness denotes an individual's overall ability to pass copies of his genes on to successive generations. For example, a woman who rears six healthy offspring has greater fitness than one who rears just two.

An inevitable consequence of variation, inheritance, and differential reproduction is that, over time, the frequency of traits that render individuals better able to survive and reproduce in their present environment increases. As a result, descendant generations in a population resemble most closely the members of ancestral populations that were able to reproduce most effectively. This is the process of natural selection.

Ecological and Ethological Approaches to the Study of Behavior

The natural history approach of Darwin and his predecessors gradually evolved into the twin sciences of animal ecology, the study of the interactions between an animal and its environment, and ethology, the biological study of animal Behavior. The roots of ethology can be traced to the late 19th and early 20th centuries, when scientists from several countries began exploring the Behaviors of selected vertebrate species: dogs by the Russian physiologist Ivan Pavlov; rodents by American psychologists John B. Watson, Edward Tolman, and Karl Lashley; birds by American psychologist B.F. Skinner; and primates by German American psychologist Wolfgang Köhler and American psychologist Robert Yerkes. The studies were carried out in laboratories, in the case of dogs, rodents and pigeons, or in artificial colonies and laboratories, in the case of primates. These studies were oriented toward psychological and physiological questions rather than ecological or evolutionary ones.

Ivan Petrovich Pavlov.

It was not until the 1930s that field naturalists—such as English biologist Julian Huxley, Austrian zoologist Konrad Lorenz, and Dutch-born British zoologist and ethologist Nikolaas Tinbergen studying birds and Austrian zoologist Karl von Frisch and American entomologist William Morton Wheeler examining insects—gained prominence and returned to broadly biological studies of animal Behavior. These individuals, the founders of ethology, had direct experience with the richness of the behavioral repertoires of animals living in their natural surroundings. Their "return to nature" approach was, to a large extent, a reaction against the tendency prevalent among psychologists to study just a few behavioral phenomena observed in a handful of species that were kept in impoverished laboratory environments.

The goal of the psychologists was to formulate behavioral hypotheses that claimed to have general applications (e.g., about learning as a single, all-purpose phenomenon). Later they would proceed using a deductive approach by testing their hypotheses through experimentation on captive animals. In contrast, the ethologists advocated an inductive approach, one that begins with observing and describing what animals do and then proceeds to address a general question: Why do these animals behave as they do? By this they meant "How do the specific Behaviors of these animals lead to differential reproduction?" Since its birth in the 1930s, the ethological approach—which stresses the direct observation of a broad array of animal species in nature, embraces the vast variety of Behaviors found in the animal kingdom, and commits to investigating Behavior from a broad biological perspective—has proved highly effective.

One of Tinbergen's most important contributions to the study of animal Behavior was to stress that ethology is like any other branch of biology, in that a comprehensive study of any Behavior must address four categories of questions, which today are called "levels of analysis," including causation, ontogeny, function, and evolutionary history. Although each of these four approaches requires a different kind of scientific investigation, all contribute to solving the enduring puzzle of how and why animals, including humans, behave as they do. A familiar example of animal Behavior—a dog wagging its tail—serves to illustrate the levels of analysis framework. When a dog senses the approach of a companion (dog or human), it stands still, fixates on the approaching individual, raises its tail, and begins swishing it from side to side. Why does this dog wag its tail? To answer this general question, four specific questions must be addressed.

With respect to causation, the question becomes: What makes the Behavior happen? To answer this question, it becomes important to identify the physiological and cognitive mechanisms that underlie the tail-wagging Behavior. For example, the way the dog's hormonal system adjusts its responsiveness to stimuli, how the dog's nervous system transmits signals from its brain to its tail, and how the dog's skeletal-muscular system generates tail movements need to be understood. Causation can also be addressed from the perspective of cognitive processes (that is, knowing how the dog processes information when greeting a companion with tail wagging). This perspective includes determining how the dog senses the approach of another individual, how it recognizes that individual as a friend, and how it decides to wag its tail. The dog's possible intentions (for example, receiving a pat on the head), feelings, and awareness of self become the focus of the investigation.

With respect to ontogeny, the question becomes: How does the dog's tail-wagging Behavior develop? The focus here is on investigating the underlying developmental mechanisms that lead to the occurrence of the Behavior. The answer derives from understanding how the sensory-motor mechanisms producing the Behavior are shaped as the dog matures from a puppy into a functional adult animal. Both internal and external factors can shape the behavioral machinery, so understanding the development of the dog's tail-wagging Behavior requires investigating the influence of the dog's genes and its experiences.

With respect to function: How does the dog's tail-wagging Behavior contribute to genetic success? The focus of this question is rooted in the subfield called behavioral ecology; the answer requires investigating the effects of tail wagging on the dog's survival and reproduction (that is, determining how the tail-wagging Behavior helps the dog survive to adulthood, mate, and rear young in order to perpetuate its genes).

Lastly, with respect to evolutionary history, the question becomes: How did tail-wagging Behavior evolve from its ancestral form to its present form? To address this question, scientists must hypothesize evolutionary antecedent Behaviors in ancestral species and attempt to reconstruct the sequence of events over evolutionary time that led from the origin of the trait to the one observed today. For example, an antecedent Behavior to tail wagging by dogs might be tail-raising and tail-vibrating Behaviors in ancestral wolves. Perhaps when a prey animal was sighted, such Behaviors were used to signal other pack members that a chase was about to begin.

Both the biological and the physical sciences seek explanations of natural phenomena in physicochemical terms. The biological sciences (which include the study of Behavior), however, have an extra dimension relative to the physical sciences. In biology, physicochemical explanations are addressed by Tinbergen's questions on causation and ontogeny, which taken together are known as "proximate" causes. The extra dimension of biology seeks explanations of biological phenomena in terms of function and evolutionary history, which together are known as "ultimate" causes. In biology, it is legitimate to ask questions concerning the use of this life process today (its function) and how it came to be over geologic time (its evolutionary history). More specifically, the words use and came to be are applied in special ways, namely "promoting genetic success" and "evolved by means of natural selection." In physics and chemistry, these types of questions are out of bounds. For example, questions concerning the use of the movements of a dog's tail are reasonable, whereas questions regarding the use of the movements of an ocean's tides are more metaphysical.

Causation

Sensory-motor Mechanisms

Honeybee (Apis mellifera).

At this level of analysis, questions concern the physiological machinery underlying an animal's Behavior. Behavior is explained in terms of the firings of the neural circuits between reception of the stimuli (sensory input) and movements of the muscles (motor output). Consider, for example, a worker honeybee (Apis mellifera) flying back to her hive from a field of flowers several kilometres away. The sensory processes the bee employs, the neural computations she performs, and the patterns of muscular activity she uses to make her way home constitute some of the mechanisms underlying the insect's impressive feat of homing. In the course of exploring these mechanisms and those underlying other forms of animal Behavior, physiologists have learned an important lesson regarding the mechanisms underlying Behavior: they are special-purpose adaptations tailored to the particular problems faced by an animal, but they are not all-purpose solutions to general problems faced by all animals. Linked to this lesson is the realization that the physiology of a species will have limitations and biases that reflect individuals' need to deal only with certain behavioral problems and only in specific ecological contexts. In Behavior, as in morphology, an animal's capabilities are matched to its expected environmental requirements, because the process of natural selection shapes organisms as if it were always addressing the question of how much adaptation is enough.

Consider first the sensory abilities of animals. All actions (such as body movements, detection of objects of interest, or learning from others in a social group) begin with the acquisition of information. Thus, an animal's sense organs are exceedingly important to its Behavior. They constitute a set of monitoring instruments with which the animal gathers information about itself and its environment. Each sense organ is selective, responding only to one particular form of energy; an instrument that responds indiscriminately to multiple forms of energy would be rather useless and similar to having none at all. The particular form of energy to which a sense organ responds determines its sensory modality. Three broad categories of sensory modalities are familiar to humans: chemoreception (exemplified by the senses of taste and smell but also including specialized receptors for pheromones and other behaviorally important molecules), mechanoreception (the basis for touch, hearing, balance, and many other senses, such as joint position), and photoreception (light sensitivity, including form and colour vision).

The capabilities of an animal's sense organs differ depending on the behavioral and ecological constraints of the species. In recognition of this fact and of the equally important fact that animals perceive their environments differently than do humans, ethologists have adopted the word Umwelt, a German word for environment, to denote an organism's unique sensory world. The umwelt of a male yellow fever mosquito (Aedes aegypti), for example, differs sharply from that of a human. Whereas the human auditory system hears sounds over a wide range of frequencies, from 20 to about 20,000 Hz, the male mosquito's hearing apparatus has been tuned narrowly to hear only sounds around 380 Hz. Despite its apparent limitations, a male mosquito's auditory system serves him perfectly well, for the only sound he must detect is the enchanting wing-tone whine of a female mosquito hovering nearby, a sound all too familiar to anyone who lingers outdoors on a midsummer's evening.

Pit vipers, colubrid snakes from the subfamily Crotalinae, which include the well-known rattlesnakes, provide another example of how the umwelt of a species serves its own ecological needs. Pit vipers possess directionally sensitive infrared detectors with which they can scan their environment while stalking mammalian prey, such as mice (Mus) and kangaroo rats (Dipodomys), in the dark. A forward-facing sensory pit, located on each side of the snake's head between the eye and the nostril, serves as the animal's heat-sensing organ. Each pit is about 1 to 5 mm (about 0.04 to 0.2 inch) deep. A thin membrane, which is extensively innervated and exquisitely sensitive to temperature increases, stretches from wall to wall inside the pit organ, where it functions like the film in a pinhole camera, registering any nearby source of infrared energy.

Human umwelt is not without its own limits and biases. Human eyes do not see the flashy advertisements to insects that flowers produce by reflecting ultraviolet light, and human ears do not hear the infrasonic calls of elephants or the ultrasonic sounds of bats. Furthermore, human noses are limited relative to those of many other mammals. Moreover, humans completely lack the sense organs for the detection of electric fields or of Earth's geomagnetic field. Sense organs for the former occur in various species of electric fishes (such as electric eels and electric catfish), which use their sensitivity to electric fields for orientation, communication, and prey detection in murky jungle streams, while the latter exist in certain birds and insects, including homing pigeons and honeybees, which use them to navigate back to the home loft or hive. At the same time, unlike most animals, humans are endowed with superb visual acuity and colour vision as a result of having evolved large, high-performance, single-lens eyes.

Each species' nervous system is an assemblage of special-purpose devices with species-specific and sometimes sex-specific capabilities. These capabilities become even more apparent when investigating how animals use their sense organs to acquire information for solving behavioral problems, such as territory defense or prey capture. Although an animal may possess diverse sensory organs that enable it to receive a great deal of information about the environment, in performing a particular behavioral task, it often responds to a rather small portion of the stimuli perceived. Moreover, only a subset of available stimuli reliably provides the information needed to perform a particular task. Ethologists call the crucial stimuli in any particular behavioral context "sign stimuli."

A classic example of sign stimuli comes from the Behavior of male three-spined sticklebacks (Gasterosteus aculeatus) when these fish defend their mating territories in the springtime against intrusions from rival male sticklebacks. The males differ from all other objects and forms of life in their environment in a special way: they possess an intensely red throat and belly, which serve as signals to females and other males of their health and vigour. Experiments using models of other fish species have shown that the red colour is the paramount stimulus by which a territory-holding male detects an intruder. Models that accurately imitated sticklebacks but lacked the red markings were seldom attacked, whereas models that possessed a red belly but lacked many of the other characteristics of the sticklebacks, or even of fish in general, were vigorously attacked.

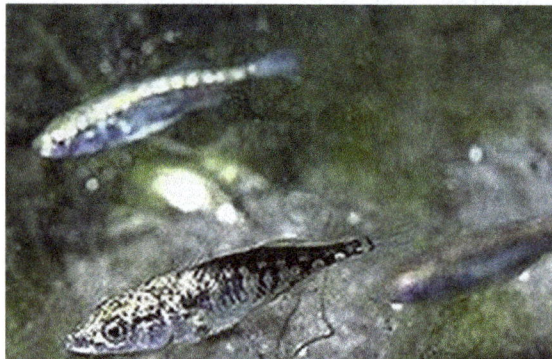

Three-spined stickleback (Gasterosteus aculeatus).

Similarly, the brain cells of some toads (Bufo) are tuned to pick out those features of the environment that reliably match the toads' natural prey items (such as earthworms). Experiments were conducted in which a hungry toad was presented with cardboard models moving horizontally around the individual at a constant distance and angular velocity. The research revealed that just two stimuli, the elongation of the object (that is, making the cardboard model longer to increase resemblance to prey) and movement in the direction of the elongation, were sufficient to initiate the toad's prey-catching Behavior. Subsequently, the toad jerked its head after the moving model in order to place it in its frontal visual field. Other stimuli, such as the colour of the model and its velocity of movement, did not influence the toad's ability to distinguish worms from non-worms, even though toads possess good colour and form vision. Even the broadly tuned human sensory system operates in a highly selective, yet adaptive, manner. For instance, a person hunting white-tailed deer seeks the prey almost exclusively by watching closely for deerlike movements amid the stationary trees of a forest, not by straining to sense the deer's shape, smell, or sound.

As with sensory systems, the neural mechanisms by which animals compute solutions to behavioral problems have not evolved to function as general-purpose computers. Rather, the central nervous system (that is, the brain and spinal cord of a vertebrate or one of the segmental ganglia of an invertebrate) performs specific computations associated with the particular ecological challenges that individuals face in their environment. A helpful illustration of this point is the startle response of goldfish (Carassius auratus). If a hungry predatory fish strikes from the side, the goldfish executes a brisk swivelling movement that propels its body sideways by about one body length to dodge the predator's attack. How does the goldfish's central nervous system process information from the sense organs to instantaneously decide the correct direction (right or left) to move? The key neural element in the startle response of the goldfish is a single bilateral pair of neurons, called the Mauthner neurons, located in the goldfish's hindbrain. Each neuron on the left or right receives input from the lateral line system (a row of small pressure sensors that are triggered by the disturbances caused by nearby moving objects) located on the left or right side of the goldfish's body. Each neuron sends output to neurons that activate the musculature on the opposite side of the body. There is strong, mutual inhibition between the left and right Mauthner neurons; should the left one fire in response to a mechanical stimulus from the left side of the body, for example, the right one is inactivated. Inactivation prevents it from interfering with the crucial, initial contractions of the trunk muscles on the goldfish's right side. The net effect is that 20 milliseconds after sensing danger the goldfish assumes a C-like shape with the head and tail bent to the same side and away from the attacker. This reaction is followed 20 milliseconds later by muscle contractions on the other side of the body so that the tail straightens and the fish propels itself sideways, away from the danger. Thus, the two Mauthner neurons of the goldfish's nervous system function exquisitely for processing information regarding predator attacks, and solving this crucial behavioral problem appears to be the only task that they perform.

Small-brained creatures, such as fishes, are not the only species whose nervous systems have evolved to solve tasks in a limited—but ecologically sufficient—way that turns difficult problems of computation into more tractable ones. For example, take the task of a human computing an interception course with a flying object, such as when a baseball player runs to catch a fly ball. In principle, the task could be solved with a set of differential equations based on the observed curvature and acceleration of the ball. What happens instead, evidently, is that the fielder finds a running path that maintains a linear optical trajectory for the ball. In other words, the player adjusts the speed and direction of his movement over the baseball field so that the trajectory of the ball appears to be straight. Unlike the more complicated differential equation approach, the linear trajectory approach does not tell the player when or where the ball will land. Consequently, the player cannot run to the point where the ball will fall and wait for it. If he did, complicating factors such as wind gusts diverting the ball might mean that he would end up in the wrong place. Instead, the player simply keeps his body on a course that will ensure interception.

Once an animal has received information about the world from its sense organs and has computed a solution to whatever behavioral problem it currently faces, it responds with a coordinated set of movements—that is, a Behavior. Any particular movement reflects the patterned activity of a specific set of muscles that work on the skeletal structures to which they are attached. The activity of these muscles is controlled by a specific set of motor neurons that in turn are controlled by sets of interneurons connected to the animal's brain. Thus, a given Behavior is ultimately the result of a specific pattern of neural activity.

Sometimes neural control takes the form of a simple sensory reflex, in which the activity in the motor neurons is triggered by sensory neurons. This activity can be achieved directly or via one or two interneurons. Other times, as in the case of rhythmic Behavior (such as with birds flying or insects walking), a central pattern generator located in the central nervous system produces rhythms of activity in the motor neurons. Central pattern generators do not depend on sensory feedback. Feedback, however, commonly occurs to modulate and reset the rhythm of the motor output after a disturbance to the animal's Behavior, as in the case of air turbulence disrupting the wing movements of a flying bird.

Most commonly, the neural control of Behavior takes the form of a motor command in which the initiation and modulation of activity in the motor neurons is produced by interneurons descending from the animal's brain. The animal's brain is where inputs from multiple sensory modalities are integrated. In this way, a sophisticated tuning of the animal's Behavior in relation to its internal condition and its external circumstances can occur. Often the control of an animal's movements involves an intricate synthesis of all three forms of neural control: patterned neural activity, simple sensory reflex, and motor command. As in all aspects of behavioral physiology, an immense diversity exists among animal species and Behavior patterns in the way the components of behavioral machinery have been linked over time by natural selection.

Cognitive Mechanisms

Cognitive psychology proposes yet another way to study the causal mechanisms of animal Behavior. The aim of cognitive psychology is to explain an animal's Behavior in terms of its mental organization for information processing (that is, how the animal acquires, stores, and acts on information present in its world). By studying cognitive mechanisms of an animal, one may study how the animal perceives, learns, memorizes, and makes decisions.

Consider, for example, crows (Corvus brachyrhynchos) that crack walnuts open by dropping them from heights of 5 to 10 metres (about 16 to 33 feet) or more onto rocks, roads, or sidewalks. The birds generally avoid dropping the nuts onto soil, where they would be unlikely to break open. Remarkably, the crows can discriminate between black and English walnuts, for they drop the harder black walnuts from greater heights. In addition, when a crow drops a nut, it takes into account the likelihood that a fellow crow might steal the contents before it can be retrieved. If fewer competing crows are perched nearby, the crow carries a nut higher into the air before releasing it. Thus, numerous processes of perception, learning, and decision-making activity underlie the crows' nut-cracking Behavior.

Each of these processes may be analyzed. For example, how do crows judge the height from which to drop nuts? Do they have to learn to adjust the dropping height in relation to the type of walnut? When faced with the conflicting conditions of having a hard-shelled black walnut and seeing a number of other crows nearby, how do they decide what drop height to use?

Until the 1970s, students of animal cognition eschewed speculation about the unobservable processing of information, limiting themselves to explaining Behaviors in terms of quantifiable relationships between stimuli and responses. Today, however, they make use of Behavior as a window into how an animal's nervous system processes information. Students of cognition also emphasize the investigation of Behaviors in which the animal does not simply respond to immediate stimuli but relies on stored representations of objects and events. For some investigators, mental representations of the environment are the essence of cognition. According to this view, known as the computational-representational approach, the experience of an animal results in the formation in the brain of isomorphisms between brain processes and events in the world. The brain then performs computations on these representations that are ultimately converted to behavioral outputs. For example, a bird assessing the availability of berries on a bush might store information about the time at which it finds each berry as it searches the bush. It might then convert this information, through a brain process equivalent to division, into a representation of the rate of berry collection.

It is possible, however, that the computational-representational approach exaggerates the richness and detail of animals' representations and the complexity of the brain processes operating on them. A good illustration comes from studies of the mechanisms by which ants (Cataglyphis fortis) living in the Sahara desert navigate home after conducting a circuitous search for food (mainly dead insects). Such a search can take these ants 100 metres (about 330 feet) or more (equivalent to 10,000 body lengths) from the entrance of their underground nest. To get back home, the ants rely on landmarks as visual signposts to show the way. Originally, it was assumed that these ants and other insects that orient using landmarks are able to store their knowledge of the nest environs in maplike internal representations called "cognitive maps." Doing so would give an ant tremendous flexibility in homing: equipped with a bird's-eye knowledge of the terrain over which it travels, an ant could return even from points where it had never before been. The mental representation used by these ants in landmark guidance is, however, actually somewhat simpler. Experiments have revealed that each ant stores a two-dimensional visual template—a kind of snapshot—of the landmark array it saw when it left its nest. When returning to its nest, the ant moves so as to match the current visual image as closely as possible with the memorized template. The snapshot-matching mechanism, unlike the cognitive-map one, enables an ant to steer its way home only from points it has recently visited, as opposed to novel sites to which it might be displaced by an experimenter. Although this mental mechanism provides a less complete and less flexible solution to the problem of finding home, it is entirely sufficient for the problems that desert ants routinely face.

An unseen and therefore largely unappreciated aspect of Behavior is the use of decision-making rules or "Darwinian algorithms." Organisms rely on these rules to process information from their physical and social environments and result in particular behavioral outputs that guide key behavioral and life-history decisions. Darwinian algorithms are made up of the sensory and cognitive processes that perceive and prioritize cues within an individual's perceptual range. These inputs are then translated into motor outputs. A Darwinian algorithm may involve a stimulus threshold (such as "when the day-length exceeds 10 hours, migrate north") or may depend on the occurrence of a cue that is normally associated with a fitness-enhancing outcome (such as "build nests in dense vegetation where chick survival is predictably high"). Darwinian algorithms are shaped through evolutionary time by the specific selective regime of each population. Which cues are relied upon depends on the certainty with which a cue can be recognized, the reliability of the relationship between the cue and the anticipated environmental outcome, and the fitness benefits of making a correct decision versus the costs of making an incorrect decision. In general, Darwinian algorithms underlying behavioral and life-history decisions are only as complex as is necessary to yield adaptive outcomes under a species' normal environmental circumstances but not so complex as to cover all experimentally or anthropogenically induced contingencies.

An intriguing question in the study of animal cognition is the role of consciousness. Humans easily distinguish between merely responding to objects and being conscious of them. For example, while driving along a highway deep in thought or conversation, the driver may suddenly realize that he has not been conscious of the road for the past several miles. Indeed, it is well documented that humans can effectively perceive, memorize, process, and even act on objects and events without the kind of awareness that underlies a verbal report of consciousness. It is possible, therefore, that the Behavior of animals occurs without conscious awareness. However, given that humans have consciousness, it seems reasonable to suppose that individuals in other species, especially social species (such as primates), also experience at least a rudimentary form of consciousness. To think otherwise would be to presume an evolutionary discontinuity between humans and all other forms of life. Thus, the possibility that at least some of the Behavior of animals is accompanied by conscious thinking seems reasonable.

Although most students of animal Behavior accept the idea that animal consciousness is a likely possibility, some argue that it is not yet possible to know whether any particular animal experiences consciousness because it is a private, subjective, and, ultimately, unknowable state. In contrast, cognitive ethologists (a separate group of animal Behaviorists), most notably American biophysicist and animal Behaviorist Donald Griffin, argue that animals are undoubtedly conscious, since individuals from a wide variety of species behave with apparent intentions of achieving certain goals. For example, chimpanzees (Pan troglodytes) stalking a monkey high above them in the treetops will distribute themselves among the trees that would otherwise provide the monkey with an escape route and attack the creature simultaneously. Similarly, groups of female lions (Panthera leo) fan out widely and then coordinate their attacks on ungulate prey. In another example, a raven (Corvus corax), when presented with the novel situation of a meat morsel dangling from a long string tied to a perch, will study the situation briefly before it acts. Subsequently, the raven will quickly procure the meat by repeatedly pulling up a length of the string with its beak and clamping each length pulled up with its feet while sitting on the perch. Studies of the states and mechanisms of animal consciousness represent important frontiers of future research.

Ontogeny

Just as a thorough understanding of an animal's morphology requires knowledge of how it develops before it hatches from an egg or emerges from its mother's womb, a complete understanding of an animal's Behavior requires knowledge of the animal's development during its lifetime. To gain this knowledge, one asks how the individual's genes and its experiences cause it to behave as it does. The ontogeny of Behavior is a subject which arouses considerable interest, perhaps because of the seeming contrast between humans and other animals in how behavioral skills are acquired. Whereas humans extensively adjust their Behavior based on experience (that is, through the process of learning), the Behavior of many animal species seems to be automatic, as if it were pre-programmed. And yet, if there really were a difference between humans and other animals in how Behavior develops, it would certainly be one of degree, not of kind.

Behavioral development is a field of study in which there have been intense clashes of opinion. Prior to the 1960s there existed a profound disagreement between European (particularly German) ethologists and American psychologists regarding methods and interpretations of such studies. The ethologists described many examples of animals showing complex Behavior patterns in response to particular stimuli under circumstances that seemed to preclude the opportunity for learning. Indeed, learning (based on external influences) was contrasted with genetic control of Behavior (based on internal influences). Austrian zoologist Konrad Lorenz, who won a Nobel Prize for his ethological studies, went so far as to classify Behavior patterns into two distinct categories: acquired and innate.

Regarding the latter, adult herring gulls (Larus argentatus) have a red spot on the lower tip of their bill. When these birds have food for their chicks, the adults point their bill downward while waving it slowly back and forth in front of the young. Newly hatched chicks will accurately peck at the red spot on the parent bird's bill, suggesting that a herring gull chick possesses innate (that is, genetically based) knowledge of where to peck for food. Ethologists termed pecking Behavior a "fixed action pattern" to indicate that it was performed automatically and correctly the first time it was elicited, apparently regardless of the animal's experience.

The psychologists, in contrast, assumed that experiences with the environment (that is, learning processes) were the main, or even exclusive, determinants of ontogeny. Accordingly, they saw nothing in the pecking Behavior of herring gull chicks that could not be explained by learning while still in the egg, conditioning, or by trial-and-error learning. For example, chicks might "learn" to peck before hatching as a result of the rhythmic beating of their heart, or they might have a pecking reflex and simply learn to associate a food reward with pecking at the parent's bill. Moreover, a chick's pecking accuracy improves with age, and after about two days it requires, in addition to the red spot, the complete configuration of an adult's head and bill to elicit pecking.

What the acquired-innate dichotomy obscured is that learning is possible only after the animal has already been steered by its genes to develop its Behavior in a certain way. An animal may well learn, but which experiences are important to the development of its Behavior depend on those that have promoted the genetic success of its ancestors. Reciprocally, whatever experiences an individual already has had can influence how its genes are activated and thus can affect their subsequent role in shaping its Behavior. Modern animal Behaviorists see the stark dichotomy of acquired versus innate as far too simplistic; no Behavior is either strictly innate or entirely learned. Rather, all Behaviors are the result of a complex interaction between genes and the environment.

Behavioral Genetics

The evidence is now compelling that genes influence Behavior in all animals, including humans. Indeed, an increasing share of biomedical research is devoted to the hunt for genes involved in human behavioral maladies such as alcoholism, obesity, schizophrenia, and Alzheimer disease. Often these studies are pursued using animal models with subjects that include mice, rats, and dogs with behavioral symptoms resembling those of humans. It is, therefore, unfortunate that the idea that genes affect Behavior is the subject of much heated and confused discussion. The principal point of confusion arises from equating genetic influence on Behavior with genetic determination of Behavior. To do so is to mistakenly believe that identifying genes "for" a Behavior implies that the gene controls, fully and inevitably, this Behavior. In actuality, to say that there are genes "for" a particular Behavior means only that within a population of individuals there exists genetic variation underlying some of the differences in this specific Behavior. To cite an example involving a morphological trait, the statement that there are genes for coat colour in guinea pigs (Cavia porcellus) or horses (Equus caballus) means that genetic variation in the guinea pig or horse population is responsible for some of the variation in coat colour.

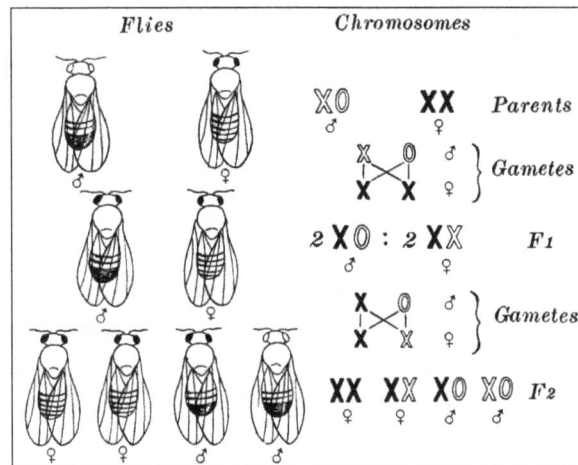

Sex-linked inheritance of white eyes in Drosophila flies.

Furthermore, identifying a gene that influences a Behavior does not imply that the Behavior is inevitable; there is considerable variation among Behaviors in the relative importance of the individual's genetic constitution and its environment to the expression of the Behavior. Occasionally, the possession of a particular form of a gene does consistently result in the individual having a particular form of a Behavior; more frequently, however, the form of the Behavior is due to a complex interaction between genes and environment.

The strength of the influence of genes on a particular Behavior is quantified by a genetic measure called "heritability." Heritability is defined as the fraction of the total variation in a trait among individuals in a population that is attributable to the genetic variation among those individuals. The remaining source of the variation is, of course, the environment. Values of heritability range between zero and one. The smaller the environmental variation experienced by the individuals in a population, the greater will be the fraction of the total variation in the Behavior that is the result of genetic variation.

One way to measure the heritability of a behavioral trait is to determine the average values of the Behavior for the parents and offspring in a sample of families within a population and calculate the linear relationship between offspring values and parental values. The slope of this line reveals the heritability of the behavioral trait in that population. For example, the heritability of the calling Behavior that male crickets (Gryllus integer) use to attract females has been measured. In any one population, some males chirp away for many hours each night, others call for just a few hours, and still others almost never call. The heritability of calling duration for one Canadian population that was studied was 0.53. The value indicates that slightly more than half of the variation in calling duration arose because males differed genetically and slightly less than half arose from environmental differences. (For example, the more parasites a cricket had acquired, the less food he had obtained, and thus the less he might be able to call on a given night.)

The degree of genetic influence on a particular Behavior is not a fixed characteristic. Rather, heritability can vary greatly depending on how much environmental variation is experienced by individuals in the specific population being studied. Thus, regarding the calling Behavior of male crickets, if every male fed well, thereby eliminating several environmental influences on calling, the numerical value of heritability would be considerably higher.

Numerous studies involving diverse species, including humans, have detected some level of heritability for every trait that has ever been examined. For example, the mean value of heritability for morphological traits, such as body and wing length, is 0.46; for life history traits, such as fecundity and life span, is 0.26; and for behavioral traits, such as calling duration and fighting stamina, is 0.30. Thus, the genetic influence on the characteristics of individual animals falls generally between 30 and 50 percent for most traits.

Instinctive Learning

An animal adjusts its Behavior based on experience—that is, it learns—when experience at one time provides information that will be useful at a later time. Viewed in this light, learning is seen as a tool for survival and reproduction because it helps an animal to adjust its Behavior to the particular state of its environment. An animal needs to know such things as what food is good to eat, when and where to find it, whom to avoid and approach, with whom to mate, and how to find its way home. When these things are not genetically preprogrammed—because they depend on the particular circumstances of an individual's time and place—the animal must learn them.

Consider, for example, a female digger wasp called the bee wolf (Philanthus triangulum) who has finished excavating a tunnel in a sandy bank. She then digs a small outpocket where one of her young will develop, and she stocks this cell with worker honeybees (Apis mellifera), which she has paralyzed by stinging and which will serve to provision her young. After laying an egg on one of the bees, she closes off the cell with sand and starts work on a new cell. To provision the cell, she must fly out to hunt more honeybees; however, after crawling out of her nest burrow, closing its entrance hole, and launching into flight, she does not immediately depart the area. Instead, she hovers just over her nest site, inspecting the ground and flying in wider and wider arcs to scan an ever-increasing area. During this elaborate departure flight, the wasp memorizes the specific configuration of landmarks—sticks, tufts of grass, and trees—around her burrow. Later, when she returns, she will use the information to pinpoint her nest's location. Her genes cannot provide her with knowledge of the landmark array around her nest, so she must learn it.

One of the clearest indications of the falseness of the old dichotomy between innate and learned Behavior is the fact that in most cases animals are genetically predisposed to acquire only specific information in developing their Behavior. One might say that most of the learning performed by animals is instinctive learning. This phenomenon is conspicuous in the flower-learning Behavior of honeybees (A. mellifera). Since at least the time of the Greek philosopher and scientist Aristotle, it has been known that worker bees show "flower constancy," a specialization by individual bees on a single species of flower. Flower constancy occurs in spite of the fact that honeybees are generalist foragers capable of exploiting many flower species. The flowers have much to gain from bees that remain faithful to them; specialist bees will be carrying the appropriate species of pollen. Therefore, the colours and odours of flowers probably evolved as conspicuous signals for the bees to learn. In turn, specialization benefits the bees by reducing flower-handling time and facilitating the collection of nectar.

Early in the 20th century, Austrian biologist Karl von Frisch demonstrated experimentally that honeybees are able to learn and distinguish a single floral odour from among at least 700 others. In addition, he found that they could distinguish colour from yellow into the ultraviolet across the electromagnetic spectrum. One striking feature of this type of colour and odour learning is the rigid programming of the timing. Research has revealed that a bee learns the flower's colour only during the final few seconds before beginning to feed, and odour learning occurs during feeding. It is as if bees possess a set of switches that turn colour and odour learning on and off at specific times during the foraging process. The time course of this learning program is highly adaptive, being restricted to times when a bee is alighted on a rewarding flower. In this manner, its learning is focused on the colour and odour of the flowers of this rewarding species rather than on the hues and scents of any nearby flowers of unrewarding species.

Is this machine like learning of bees fundamentally different from the learning processes in vertebrates? Until the mid-1960s, psychologists generally believed so. Studying mainly birds and mammals, they developed an approach known as "general process learning theory," which attempted to account for learning with a single set of principles, namely unconstrained "associative learning" as studied in instrumental (operant) conditioning and classical (Pavlovian) conditioning. Associative learning is said to occur when an animal changes its Behavior upon forming an association between an environmental event and its own response to the event. In operant conditioning, the animal learns to associate a voluntary activity with specific consequences. In classical conditioning, the animal learns to associate a novel (conditioned) stimulus with a familiar (unconditioned) one. For example, in his study of classical conditioning, Russian physiologist Ivan Petrovich Pavlov demonstrated that by consistently exposing a dog to a particular sound (novel stimulus) and simultaneously placing meat powder (familiar stimulus) in its mouth the dog could be made to salivate upon hearing the sound even without the meat stimulus. Initially, salivation was the unconditioned response, whereas the food stimulus was the unconditioned stimulus. Once the dog learned to associate the sound stimulus with the food stimulus, salivation became the conditioned stimulus to sound—that is, a stimulus that previously did not trigger a response.

The popularity of general process learning theory peaked in the 1940s and '50s. In the mid-1960s, however, American psychologist John Garcia discovered several puzzling phenomena that indicated adaptive limits on learning and contradicted the supposedly general principles of conditioning. One of the most important of these anomalies was flavour aversion learning. When rats (Rattus norvegicus) and many other vertebrates, including humans, sample a flavour and later become ill, they learn to avoid consuming that flavour in the future. This phenomenon has two remarkable properties. First, it occurs despite delays of several hours between experiencing the flavour (the conditioned stimulus, or CS, in the Pavlovian conditioning paradigm) and experiencing the illness (the unconditioned stimulus, or US); it does not require the brief delay specified by the general principles of conditioning. Second, in rats, learning with the US being illness is limited to flavours. This response was revealed in an experiment in which rats experienced a flash of light and the sound of a buzzer each time they took a drink from a tube of flavoured water (hence "bright noisy tasty water" became the CS). Some of the rats were made ill (nauseous) after drinking (hence illness became the US for them), whereas others were shocked through the feet shortly after they began drinking (hence pain became the US for them). After conditioning, the rats were tested with the noise plus the light alone or with the flavour alone. Those rats that had been made ill avoided drinking only the "tasty water," whereas the rats that had been shocked avoided drinking only the "bright noisy water." In other words, the rats could learn to associate a taste with an illness but not a visual and auditory stimulus. Conversely, the rats could learn to associate a visual and auditory stimulus, but not a taste, with pain.

These findings attracted tremendous skepticism when they were first reported because both the long delay between CS and US and the CS-US specificity contradicted the idea of general laws of learning. Both findings, however, make considerable sense in light of the problems faced by rats living in nature. If they consume a new food and become ill even hours later, they will not eat the food again and thus not suffer the illness associated with the food. Moreover, it is adaptive that rats learn to associate a taste cue, not an auditory or visual cue, with illness-causing food because rats discriminate foods best using chemical cues sensed by taste, olfaction, or both. In contrast, something that causes pain is best recognized from a safe distance. Therefore, it is adaptive that rats learn to associate auditory and visual cues with painful experiences. Thus, these "anomalies" for general process learning theory can be understood by considering the functions that the rats' learning has evolved to serve.

There is now compelling evidence that humans also possess adaptive predispositions in learning abilities. Consider, for example, the curious anthropological discovery made in 1926 by Finnish sociologist Edward Westermarck that arranged marriages between children that grow up together (whether biological siblings or not) are far more likely to fail than arranged marriages between individuals not raised together. The failures most often result from sexual incompatibilities. Evidently, children are genetically guided to learn to treat as siblings all individuals with whom they are raised together. And because siblings tend to avoid sexual contact, presumably due to a long evolutionary history of detrimental consequences associated with inbreeding, marriages between these individuals tend to fail.

Today it is widely recognized that the general-purpose psychological approach to learning had overlooked its biological significance and that animals possess learning mechanisms that are specialized for solving the problems they face in the natural world. This view of learning explains the psychologists' observations of the limits of learning by animals in laboratory settings. It also makes sense of ethological reports of special forms of learning, such as imprinting (that is, the rapid identification of parents by newborn animals triggered by following the first object they see moving away from them), which have been studied in naturalistic settings. To a large extent, this picture of instinctive learning has brought a constructive end to the centuries-old debate about whether "nature" (genes) or "nurture" (experiences) is the source of adaptive behavior of animals. Animals are shaped by their experiences; however, the interpretation of each experience is governed by a collection of rules (Darwinian algorithms) set by the genes in each species.

The general-purpose view of learning that prevailed during most of the 20th century was based on two assumptions: (1) the ability to learn is always beneficial, and (2) animal learning abilities are like human learning abilities, which seem to be of completely general and unlimited applicability. Neither assumption is correct.

First, there are costs as well as benefits to learning, so learning abilities will be beneficial, and favoured by natural selection, only when the benefits outweigh the costs. The costs include those involved in building and maintaining the required neural circuitry and also the time and mistakes involved in learning while the animal is fine-tuning its Behavior to the current or likely future state of its environment. When learning is a matter of life or death—as in geese (Anser and Branta), sheep (Ovis), and antelopes (family Antilopinae), where newborn young must keep up with mobile parents—the advantage of rapid learning (that is, staying together) and the danger of slow learning (that is, lagging behind) are both extremely high. By considering both the fitness costs and the benefits of different forms of learning, one can readily appreciate the reasons why imprinting occurs in these species, rather than the slower process of trial-and-error learning.

Second, as described earlier, the learning abilities of animals, including humans, are not completely general; learning abilities are adaptively specialized so that, in any particular context, animals take in only the most relevant information. Late in his career, Lorenz referred to "the innate schoolmarm," a phrase that picturesquely expresses the reality that animals possess adaptive predispositions in their learning.

Function

In studying the function of a behavioral characteristic of an animal, a researcher seeks to understand how natural selection favours the Behavior. In other words, the researcher tries to identify the ecological challenges, or "selection pressures," faced by a species and then investigates how a particular behavioral trait helps individuals surmount these obstacles so that they can survive and reproduce. In short, the question being asked is: What is the behavior good for?

Until the mid-1960s, functional interpretations of animal Behavior were usually made in terms of how a behavior was "good for the species." Social Behaviors that excluded some individuals from reproducing (such as territorial defense and courtship displays) were seen as adaptations for regulating animal populations at levels that would prevent overpopulation, environmental destruction, and extinction of the species. This view was based on the observation of ecological phenomena—such as the overgrazing of grassland by cattle, leading to the starvation of the animals. American evolutionary biologist George C. Williams and British ornithologist David Lack, however, revealed the underlying theoretical problem with the view that animals behave in ways that limit their reproduction for the good of their species. Williams noted that individuals who maximize their own reproduction will have greater genetic success than those who behave in ways that limit their reproduction. Thus, over time, in subsequent generations, reproduction-reducing Behaviors will be replaced by reproduction-enhancing ones. Therefore, it has become evident that it is incorrect to interpret the behavior of animals as having evolved to function "for the good of the species." Instead, the appropriate interpretation is how a behavior has evolved for the "good of the individual."

Williams's theoretical argument was bolstered by Lack's long-term study of the reproductive Behavior of the European, or common, swift (Apus apus). At first glance, swifts appear to voluntarily restrict their own reproduction. When Lack removed the eggs laid each day from a pair's nest he discovered that the female could lay up to 72 or more eggs in a season. Yet, surprisingly, she usually lays just two or three eggs. Are chimney swifts regulating their egg production to avoid overpopulation, or does the number of eggs laid equal the number of young they can successfully rear each year? Lack answered this question by performing the experiment of adding one or two nestlings to the nests of certain pairs so that, instead of the normal two or three young, they would have to rear four or five. He then compared the reproductive success of these pairs to those that were left rearing the normal number. Lack found that the birds with four or five young were less successful (that is, rearing fewer young to fledging) than those in a control group who reared a normal-sized brood. Therefore, chimney swifts, in rearing just two or three offspring, are not withholding reproduction for the good of their species or local population; instead, they are producing as many young as they can successfully rear given a limited food supply, thereby maximizing their own reproduction.

Chimney swifts provide just one example of a pattern that has been found repeatedly by biologists studying the Behavior and reproduction of animals. They have found that individuals are "selfish," behaving in ways that benefit their own reproduction regardless of its long-term effect on the survival of their species. Sometimes, however, animals engage in apparent altruism (that is, they exhibit Behavior that increases the fitness of other individuals by engaging in activities that decrease their own reproductive success). For example, American zoologist Paul Sherman found that female Belding's ground squirrels (Spermophilus beldingi) give staccato whistles that warn nearby conspecifics of a predator's approach but also attract the predator's attention to the caller. Likewise, worker honeybees (Apis mellifera) perform suicidal attacks on intruders to defend their colony, and female lions (Panthera leo) sometimes nurse cubs that are not their own.

The key insight to understanding the evolution of such self-sacrificial Behavior was provided by British evolutionary biologist William D. Hamilton in the mid-1960s. He argued that natural selection favours genetic success, not reproductive success per se, and that individuals can pass copies of their genes on to future generations. Genes are passed from direct parentage (the rearing of offspring and grand-offspring) and by assisting the reproduction of close relatives (such as nieces and nephews), a concept referred to as "inclusive fitness" or "kin selection."

Hamilton devised a formula—now called Hamilton's rule—that specifies the conditions under which reproductive altruism evolves:

$$r \times B > C$$

where B is the benefit (in number of offspring equivalents) gained by the recipient of the altruism, C is the cost (in number of offspring equivalents) suffered by the donor while undertaking the altruistic Behavior, and r is the genetic relatedness of the altruist to the beneficiary. Relatedness is the probability that a gene in the potential altruist is shared by the potential recipient of the altruistic Behavior. Altruism can evolve in a population if a potential donor of assistance can more than make up for losing C offspring by adding to the population B offspring bearing a fraction r of its genes. For example, a female lion with a well-nourished cub gains inclusive fitness by nursing a starving cub of a full sister because the benefit to her sister (B = one offspring that would otherwise die) more than compensates for the loss to herself (C = approximately one quarter of an offspring), since the survival probability of her own, non-starving cub is only slightly reduced. Given that the average genetic relatedness (that is, r) between two full sisters is 0.5, then according to Hamilton's rule $(0.5 \times 1) > 0.25$. In essence, genes for altruism spread by promoting aid to copies of themselves.

According to this view, which was popularized by British zoologist Richard Dawkins, the most appropriate way of viewing natural selection is from a gene-selection perspective, as embodied in Hamilton's rule. Genes that are best able to guide the organisms that bear them to propagate successfully will persist and proliferate over generations. Consequently, an explanation of the function of a particular Behavior should include how the Behavior promotes the success of the genes that underlie the Behavior. Of course, since an animal's Behavior almost always promotes genetic success by helping the animal survive and reproduce its genes, investigations of behavioral function typically address the survival and reproductive value of the Behavior.

Natural Selection in Action

The most straightforward way to study the function of a Behavior is to see how natural selection operates on it under current conditions by studying differential reproduction. Often this kind of investigation can be conducted by exploiting the naturally occurring variation among individuals, such as in a particular phenotypic (observable) trait in a population. Sometimes, however, the researcher must experimentally enhance behavioral variation where too little exists in nature. The experimental approach may have the disadvantage of involving unnatural variants, but it has the advantage of revealing how differences among individuals, even in a single trait, can cause variation in reproductive fitness. Either way, a study of natural selection acting on Behavior requires that the researcher be able to observe natural populations and obtain detailed information on each individual's survival, its ability to attract a mate, its fertility, and so forth. All of this information is essential to assess an animal's success in passing on its genes.

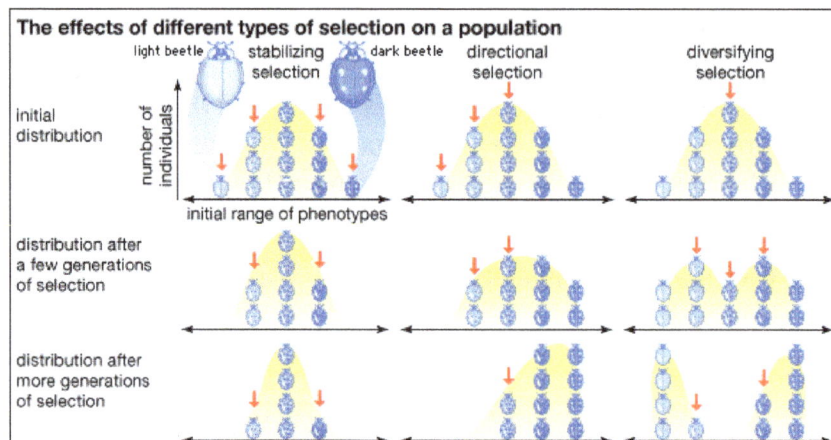

In above figure: Three types of natural selection, showing the effects of each on the distribution of phenotypes within a population. The downward arrows point to those phenotypes against which selection acts. Stabilizing selection (left column) acts against phenotypes at both extremes of the distribution, favouring the multiplication of intermediate phenotypes. Directional selection (centre column) acts against only one extreme of phenotypes, causing a shift in distribution toward the other extreme. Diversifying selection (right column) acts against intermediate phenotypes, creating a split in distribution toward each extreme.

An investigation of why male titmice, or great tits (Parus major), woodland birds of Europe, sing multiple songs serves to illustrate how a behavioral function can be studied by exploiting naturally existing variation. Each great tit male has a repertoire of one to eight songs that he uses to advertise his presence on a territory. Investigators can acquire detailed information on the breeding biology of these birds because great tits are cavity nesters that readily accept man-made nest boxes. In one experiment on a wooded estate near Oxford, Eng., English zoologist John Krebs and his colleagues installed and regularly inspected nest boxes during the breeding season. The researchers recorded the singing Behavior of each breeding male in order to determine repertoire size. They also recorded the egg-laying date, the clutch size (number of eggs), the brood size (number of young), and the fledgling weight for the nests of numerous males. It was possible to monitor the survival of each male's young to the time of its own breeding, because all the young were banded before they fledged and most fledglings returned to the same woods to breed themselves.

The researchers found that individual tits had different repertoire sizes. Males with larger repertoires had chicks that were heavier at fledging, and more of these chicks survived to breed than offspring of males with smaller repertoires. Thus male repertoire size and reproductive success were correlated. The underlying mechanism is that males with larger song repertoires were able to acquire superior territories—specifically, ones with better food. Previous studies had shown that size and survival of young tits depend on body weight at time of fledging: the bigger and heavier the fledgling, the greater its chances of survival to maturity. Thus, the function of a great tit male's singing multiple songs is to help him secure a top-quality breeding territory and mate. So why do all males not sing multiple songs? Perhaps songs are learned over time, so that only the oldest males can possess a large repertoire. Alternatively, perhaps there are costs (such as time away from foraging or increased vulnerability to predators) to singing multiple songs, and only the biggest, strongest males can sing many songs and still survive.

Direct comparisons of individuals of the same species exhibiting natural variation in Behavior is a revealing way to study behavioral function. However, when appropriate natural variations do not exist, experimental manipulations can provide the needed variation in the Behavior. The variant forms are then studied in the field to determine how well extreme forms of the Behavior do in the face of natural selection. Using this method, American biologist Thomas Seeley investigated nest site choice in a species of Southeast Asian honeybee, Apis florea. Colonies build their nests of beeswax combs amid dense foliage, suspended from the branches of bushes and understory trees. Moreover, if a colony's nest loses its cover during the dry season when many trees shed their leaves, the colony will build its new nest in another leafy site. What is the function of this Behavior of nesting in dense vegetation? Is it to prevent the nest from overheating under the strong tropical sun, or to conceal the nest from predators, or both?

To test the antipredator hypothesis, pairs of naturally occurring colonies were identified. Within each pair the vegetation around the nest of one colony, which served as the experimental unit, was removed, leaving only enough to provide shade but rendering it conspicuous to predators. The vegetation surrounding the nest of the second colony, which served as the control, was not removed. Measurements of nest site temperatures one day later revealed no significant differences between the two nests. Within one week, however, four of the seven experimental colonies had been discovered and destroyed by predators (probably monkeys and tree shrews) whereas none of the control nests had suffered any damage. Thus, it appears that A. florea colonies choose dense vegetation as nesting sites primarily to conceal their nests from predators.

Another example of a well-controlled field experiment on the function of Behavior is Dutch-born British zoologist and ethologist Nikolaas Tinbergen's pioneering study of eggshell removal by black-headed gulls (Larus ridibundus). In a matter of hours after their eggs hatch, they pick up the empty eggshells, fly off, and drop them well away from the nest. Why should a gull engage in this Behavior? One hypothesis was that the sharp edges of the shells might injure the chicks, a danger that is well known to poultry breeders. Another hypothesis was that the white insides of broken shells might attract predators, such as crows and herring gulls flying overhead, and so endanger the brood. To test the latter hypothesis, Tinbergen and his colleagues distributed single gull decoy eggs around the dunes where the black-headed gulls nest, and placed broken eggshells near some of the decoy eggs while leaving others isolated. The investigators found that the eggs near broken shells were preyed upon sooner than the isolated, less conspicuous eggs. Evidently, the removal of broken eggshells from the nest by gulls helps to maintain the camouflage of the brood, thereby reducing predation.

Adaptive Design

Many features of animal Behavior are so well suited to their function that it is impossible to imagine that they arose by chance. Echolocation by bats, the nest-building skills of weaver birds (family Ploceidae), and the alarm signals of ground squirrels all serve obvious purposes, and the mechanisms that enable them are remarkably similar to what engineers would design to achieve those ends. However, such adaptive Behaviors have no divine designer but instead have arisen through the process of natural selection.

Natural selection is an inherently optimizing process: it favours those versions of an organism's traits, including behavioral ones, which best enable the organism to propagate copies of its genes into future generations over alternative versions with lower fitness. Creating a formal optimality model is one way to infer the adaptive "design" or function of a Behavior. Using an engineering or economic model to work out the optimal behavioral solution for a given ecological problem is a way of specifying the best design out of a wide range of alternative possibilities. Therefore, if an optimality model embodies an accurate understanding of the function of a Behavior, it can predict the form of the Behavior that is observed in nature.

One of the attractions of using optimality models to test hypotheses about functional design is that these models yield quantitative predictions that can be easily tested. If a model's predictions regarding the form of a Behavior do not match reality, one knows immediately that the hypothesis expressed in the model is false. For example, foraging honeybees often return to the hive with less than a full load of nectar, and biologists initially assumed this was because a bee maximizes its rate of energy delivery to the hive. The fuller the bee, however, the slower she can fly. As a result, the transportation of a full load was assumed to depress a bee's rate of nectar collection. On the other hand, when the bees were trained to forage from an array of artificial flowers in which each flower offered a fixed amount of nectar and the time spent flying between flowers was varied to alter the duration and cost of foraging, the size of the bee's load did not maximize her net rate of energy delivery to the hive. Further analysis revealed that a bee's decision of when she would return to the hive is based on the maximization of foraging efficiency. Evidently, bees behave so as to achieve the highest foraging efficiency rather than the highest food-delivery rate to the hive.

A classic example of application of the optimality approach to understanding the adaptive design of a Behavior is a study of copulation time in the yellow dung fly (Scatophaga stercoraria) by British evolutionary biologist Geoffrey A. Parker. Shortly after cow excrement is deposited in a meadow, it is invaded by female dung flies that come to lay their eggs on the dung and by males seeking to mate with the females. Competition among the males for females is fierce. Sometimes one male succeeds in kicking a rival off a female during copulation and mounts her himself. Unfortunately for the first male, this means that some of the female's eggs will be fertilized by the second male. The longer the first male copulates, the more eggs he fertilizes, but the returns for extra copulation time diminish rapidly. How much time should a male spend copulating with a female? Should he copulate for as long as is needed to fertilize all the eggs (about 100 minutes), or should he quit earlier (or permit himself to be displaced) so that he can go search for a new female? Parker hypothesized that a male dung fly chooses a copulation time that maximizes his overall rate of egg fertilizations. He tested his hypothesis using a graphical optimality model.

Before a male dung fly that has just finished copulating with a female can copulate with a new one, he must spend on average 156 minutes searching for her. Once he has found a new female, the proportion of her eggs fertilized by him as a function of copulation time is set by female physiology, and this has been quantified as a curve based on experimentally measured values. The male cannot shorten the time necessary to find a new female or change the fertilization curve, but he can stop copulating at will. The optimal solution, assuming that his decision regarding copulation duration serves to maximize his rate of egg fertilizations, is to copulate for 41 minutes. Because the average observed copulation time, 38 minutes, is quite close to the predicted time of 41 minutes, it is clear that the Darwinian algorithm underlying a male dung fly's copulation Behavior serves to maximize his rate of egg fertilizations.

A second way of studying the adaptive design of a Behavior is what Darwin called the comparative method, which takes advantage of the thousands of "natural experiments" that have occurred over evolutionary time (that is, throughout the formation of new species and the evolution of their special characteristics). Here again, specific hypotheses regarding how natural selection has shaped a Behavior are tested. Rather than simply examining one species, behavioral researchers collect data from a number of species simultaneously. The idea is to compare the degree to which a particular Behavior occurs in each species with the degree to which the hypothesized selection pressure is part of the ecology of each species.

Australian zoologist Peter Jarman was one of the first to use the comparative method to study the diversity of mating systems, specifically among various species of African antelope. In some species, such as the dik-dik (Madoqua), individuals are solitary and cryptic; however, during mating season, they form conspicuous monogamous pairs. Others, such as the black wildebeest (Connochaetes taurinus), form enormous herds. During the breeding season, only a few males control sexual access to a group of females in a polygynous mating system. When Jarman compared these African ungulates, he found that body size, typical habitat, group size, and mating system were interrelated. Specifically, smaller species with relatively high metabolic rates (such as the dik-dik) need to consume high-quality food—such as fruits and buds in the forests—while concealing themselves from predators. Because of the sparse distribution of food and the need to remain solitary and cryptic to avoid capture by predators, the smaller species are widely dispersed, leaving no opportunity for a single male to monopolize access to many females. Consequently, small-bodied species tend to be monogamous. In contrast, the larger species graze in open plains where food is generally abundant, although seasonally variable in its geographic distribution, and they are highly visible to predators. Thus, species such as the wildebeest live in large herds that migrate with the seasons. Each individual may be hidden within the large number of other animals in the herd; however, group living creates the opportunity for one male to monopolize several females, and polygyny tends to be found in the large-bodied species. This pattern, which holds true for birds and primates as well as ungulates, supports the hypothesis that the mating system of a species is derived from selection pressures associated with food and predation. Selection pressures determine the spatial distribution of females and thus their defensibility by individual males.

Not all comparative analyses of Behavior are so broad. Some focus on just one Behavior or a morphological correlate of Behavior. Consider the case of sexual dimorphism in body size where the males of some species tend to be considerably larger than the females. It had been hypothesized that size is a key advantage in species where males must fight to defend females from rival males. To test the hypothesis that sexual dimorphism was favoured by natural selection, American evolutionary biologist Richard Alexander and his colleagues compared social structure of the breeding group in primates, ungulates, and pinnipeds with their degree of body-size dimorphism. They reported that body size is similar between males and females in species, including humans, where the breeding group typically consists of one male and one female or a few females. Male body size, however, increases compared with female body size in species that breed in groups made up of multiple males and females, and it is highest in species where a single male defends a large group of females. Evidently, male size in primates is an adaptation related to the intensity of male-male physical competition for females.

Evolutionary History of Behavior

Biologists have always been fascinated with the question of where the traits that exist today came from—that is, their evolutionary history. However, exploration of the history of Behaviors and their underlying mechanisms is exceptionally challenging. (Unfortunately, the fossil record is largely uninformative.) Only under rare circumstances, such as the discovery of a fossilized dinosaur nest topped by an adult (a situation suggestive of parental care), is there sufficient information captured in fossils to enable paleontologists to draw inferences about the origin and subsequent evolution of complex social or reproductive Behaviors. As a result, it has been necessary to develop alternative and indirect approaches to infer evolutionary histories of Behaviors.

Character Mapping

The first approach, called character mapping, begins by constructing a phylogenetic tree (that is, a depiction of the presumed relationship of a species of interest to its closest living relatives). Phylogeny refers to the evolutionary history of one or a group of interrelated species. Hypotheses regarding phylogenetic relationships often are based on similarities among existing species in morphological traits and DNA sequences. Once the phylogenetic tree is established, character states, or Behaviors (such as parental care), of extant species are attached, or "mapped," to it. Sites on the tree called ancestral nodes are drawn where changes in the Behavior of interest apparently occurs. This is accomplished by minimizing the number of character state transitions, or changes, necessary to account for all the diversity seen among the related species today. In other words, the shortest evolutionary path taken by any character from its origin to the present is considered to be the "most parsimonious" (that is, requiring the fewest changes) and, therefore, the most probable. Assuming that the Behaviors of extant species have remained the same since the last speciation event in their lineage and that the shortest evolutionary path is indeed most likely, a hypothesis can be formulated about the relative timing of the origins of various Behaviors and their subsequent loss or evolutionary modification. These assumptions are most valid for complex Behaviors whose evolution required many improbable changes rather than highly variable (plastic) Behaviors. Moreover, it is more reasonable to suppose that a complex Behavior that is shared by two or more species was present in a common ancestor than that it evolved multiple times independently.

Phylogenetic reconstructions and character mapping have been used to infer the historical trajectories of male secondary sexual characteristics and female mating preferences in several taxa, such as Central American frogs (Physalaemus) and swordtail fishes (Xiphophorus). In the frogs, electrophysiological studies of present-day species indicate that females have identical auditory preferences regardless of the acoustic characteristics of the mating calls of the males. The most parsimonious hypothesis, therefore, is that female preferences evolved first (that is, they are ancestral or older), and that male calls evolved secondarily in some species to take advantage of these preexisting preferences. In the swordtail fishes, females in species with and without swords prefer males with artificial swords attached to their caudal fins over unsworded males. The hypothesis that ancestral females possessed the preference for a swordlike structure is more parsimonious than that the preference for swords evolved multiple times independently in the lineage of each existing species.

One general problem with the character mapping approach is that the most parsimonious evolutionary pathway may not be the most likely. Evolutionary change is seldom unidirectional, so small changes in characters in one direction or the other may have occurred multiple times over the evolutionary history of a species group. A more specific problem with inferring the evolutionary history of sexually selected characters using character mapping is that it is often difficult to determine exactly what aspects of a male trait females prefer. With reference to swordtail fishes, it is unclear whether females have specific preferences for a trait (such as the sword) not possessed by the males or whether females are attracted to any tail modifications that are indicative of male viability or fertility in general (such as relatively large, brightly coloured, healthy, and vigorous males). In other words, do swordtail females really prefer sworded males per se or are they attracted to any males capable of growing brightly coloured and exaggerated tails? Recent evidence suggests the latter.

Phylogenetic Grading

A second approach to inferring evolutionary history may be referred to as "phylogenetic grading." The approach involves making detailed comparisons among extant species with respect to a particular type of Behavior and then arraying the various forms of this Behavior from least to most complex. Assuming that complexity increases over evolutionary time, simple or more "primitive" forms of a Behavior are considered ancestral. Species that exist today with a simpler form of the Behavior are not presumed to have experienced the selection pressures that propelled the evolution of more complex forms of the Behavior in other species. For example, Austrian zoologist Karl von Frisch, who decoded the "dance language" of honeybees (Apis), reportedly said:

> "We cannot believe that the bee dance of the European bees has come from heaven as it is and, since the Indian honeybees and the stingless bees there live in a more primitive social organization, we should expect some phylogenetically primitive stages of the bee dance."

According to this view, stingless bees (Melipona) might not even possess a dance language, since they live in small, less-organized colonies (that is, they are lower on the phylogenetic grade of social complexity than honeybees). Recent studies of stingless bees, however, indicate that successful foragers do in fact communicate distance, direction, height, and smell of food sources to their colony mates. In other words, stingless bees can do everything that the more "advanced" honeybees do—and more, because honeybees do not indicate food-source height. Stingless bees have a communication system that is different from, but certainly not more primitive than, the communication system of honeybees.

The phylogenetic grade approach probably appeals to investigators because of the human tendency to admire the technological advances that have occurred in human societies. So-called advanced species with complex Behaviors and social structures, however, are really no better adapted than so-called primitive species, and complexity is no guarantee of long-term success. Many species with complex Behaviors are extinct (such as the dinosaurs), and in some extant phylogenetic groups (such as bowerbirds [family Ptilonorhynchidae]) there are species living today whose ancestors probably engaged in much more complex bower-building activities. In other words, living species with simple Behavior patterns are sometimes descendant from ancestral species with more complex Behaviors, and vice versa. Consequently, it is inappropriate to view the Behavior of living species as the rungs of a ladder of complexity progressing back to simpler ancestral Behaviors. Natural selection does not inexorably build complexity but rather promotes only the complexity necessary at any given time for survival and reproductive success.

Artificial Selection

A wholly different approach to reconstructing the evolution of certain Behaviors involves the attempt to "re-create" history by imposing an artificial selection regime on a species that is closely related to the one showing the Behavior of interest. The selection that is imposed is designed to mimic what might have occurred in a past environment of the species exhibiting the focal Behavior. For instance, to show how dogs may have acquired their domesticated traits, Russian geneticist Dimitry Belyaev imposed artificial selection on a closely related but undomesticated species, the silver fox, a colour morph of the red fox (Vulpes vulpes). After capturing a group of wild foxes, he bred them in captivity. Once a month, starting when each pup was one month old, he offered food and tried to approach and pet it. When the foxes were seven to eight months old, only those that were enthusiastic about human contact were selected as breeding stock. After 40 years of this strong and consistent artificial selection for tameness, the farmed foxes behaved like house dogs, whimpering to attract attention, wagging their tails, licking handlers, and sitting in their handlers' laps. Interestingly, in addition to behavioral changes there were changes in morphology as well, including floppy ears, shortened legs and tails, tails curved upward, underbites and overbites, and novel coat patterns and colours.

Belyaev's analyses indicated that the ontogeny of the farmed foxes' social Behavior had changed: their eyes opened earlier and their fear response was initiated later, widening the window of time for social bonding. As the Behavior of the foxes evolved, changes took place in the mechanisms that regulated development, leading to shifts in the rates and timing of developmental processes such as socialization. Floppy ears, recurved tails, and bizarre colours probably are genetically correlated traits, meaning that their development is affected by the same genes that result in tameness. It is possible that the fox experiment re-created the process by which wolves (Canis lupus) became domesticated into house dogs 10,000–15,000 years ago. Moreover, the striking similarities of many of the Behaviors and physical attributes of domesticated swine (Sus domesticus), horses (Equus caballus), cows (Bos taurus), and cats (Felis catus) to those of the foxes suggest that the Behavior of all those animals followed a similar evolutionary trajectory. Domestication of those animals was the result of selection imposed by humans for tameness.

The Comparative Approach

The fourth approach to reconstructing the history of a Behavior involves studying its fitness consequences today. If a Behavior currently provides higher fitness than its alternatives, it is inferred that natural selection acting in similar antecedent environments caused its initial spread. This approach assumes that present selective pressures are similar to those that operated in the past. This assumption is reasonable because the physical and biotic environments of many organisms have remained similar for hundreds of thousands, and even millions, of years. Even if certain aspects of the environment of a species have changed recently, other aspects may have remained the same. For this approach to succeed, the only environmental aspects that matter are those to which the focal Behavior is a response.

For example, the European (or common) starling (Sturnus vulgaris) and the English (or house) sparrow (Passer domesticus) were imported to the United States during the second half of the 19th century. Certain aspects of their new environment—such as types of food and predator species—were different, whereas other environmental aspects—such as nesting sites and the birds' social environment—did not change (the latter is a product of the birds' tendencies to group with members of the same species). As a result, the birds' reproductive and communicative Behaviors closely resemble those of starlings and sparrows living in Europe today. Therefore, studies of current fitness in the new, nonnative environment would still be relevant to reconstructing the history of starling and sparrow nesting and social Behaviors (such as mate choice and parental care) although perhaps not relevant for inferring the history of the birds' foraging or antipredator Behaviors.

The current fitness approach has been used to reconstruct the history of human social Behaviors. This is largely because the other three approaches are precluded. Societies of chimpanzees (Pan troglodytes) and gorillas (Gorilla gorilla), the closest phylogenetic relatives to human beings (Homo sapiens sapiens) are so different from human societies that character mapping of Behaviors is of limited usefulness, and selection experiments on humans are considered unethical. There exist, however, alternative forms of many human social Behaviors, and these alternative forms may well give rise to fitness differences among individuals. Although there are vast differences between certain aspects of today's environments and those experienced by humanity's ancestors (as a result of technological advances), other aspects have changed very little (such as the dangers of parasites and infectious diseases, the desirability of attracting a mate, family-based social units, parental Behaviors, nepotism, and reciprocity). Therefore, the approach of studying current fitness consequences is suitable for humans.

The match between ancestral and modern environments can sometimes be improved by studying the Behavior of humans living in societies without advanced technologies. These so-called traditional societies may offer a window into the evolutionary past since it is almost certain that ancestral Homo sapiens were hunters and gatherers. Thus, by examining modern hunting and gathering societies, insights can be gained into the conditions confronted by ancestral humans and the Behavior patterns they used to survive and reproduce. Such analyses have revealed many differences in the Behaviors of humans living in various traditional societies, as well as those living in highly technological societies, suggesting that humans have evolved capacities to adjust Behavior in different environments to benefit themselves and their kin. At the same time, commonalities have emerged both within and between traditional and highly technological societies. These commonalities occur in Behaviors (such as mate choice and patterns of nepotism and reciprocity) and in parental roles. For example, greater parental solicitude toward one's own offspring than toward unrelated children, along with the avoidance of incest, is universal. A sexual division of labour in foraging also appears to be common. In many societies, women gather vegetable foods and men hunt; however, in a few other societies labour is shared or roles are reversed. Sexual differences in mate-choice criteria are also universally widespread. Women of most societies prefer older, wealthy men of high social status, whereas men in most societies prefer younger, healthy, fecund women. The implication of these commonalities is that these similarities and differences are evolutionarily ancient.

Comparative studies can yield hypotheses about the origins of Behaviors that can sometimes be tested indirectly with fossil evidence. For example, if a certain Behavior is associated with a particular morphological structure, such as an elongated tail, the appearance in the fossil record of that structure confirms the time of origin of the associated Behavior. In this manner, the approach used to develop the hypothesis regarding the evolutionary history of that Behavior is also validated.

In conclusion, there are several different ways to tackle the knotty problem of evolutionary history, but none is completely satisfying. Indeed, it seems impossible to achieve complete certainty about a Behavior's origin and evolutionary trajectory. Without rock-solid fossil evidence, the best attempts to reconstruct behavioral evolution will yield valid references, but they will not produce strong conclusions.

ETHOLOGY

Ethology is defined as the systematic and scientific study of the Behavior of animal (including human) under natural conditions.

Genetics, developmental biology, anatomy, physiology, endocrinology, neurobiology, evolution, learning and social theory are all combined into one grand subject — animal Behavior. The field of ethology is, thus, integrative in the true sense of the word.

Approaches to Ethology

The approach to ethology is based upon three foundations — the forces of natural selection, the ability of animals to learn and the power of transmitting learned information.

1. The Forces of Natural Selection

According to Darwin any trait that causes its possessor to have some sort of reproductive advantages, would be favoured by the process, which he named as natural selection. Thus, natural selection is the process whereby traits that confer the highest relative reproductive success (greatest relative fitness) on their bearers and which can be passed down across generations, increases in frequency over many generations.

To elaborate how natural selection operates in the wild, let us take the example of beak size in Galapagos finches, which is also called Darwin's finches. Two such finch species', Geospiza magnirostris and G. fortis, beak size can be utilised to elaborate the role of natural selection in animal Behavior.

The bigger of the two species is G. magnirostris, which has relatively large beak. It can crack open large and very tough fruit of the caltrop (Tribulus cistoides) much more quickly and efficiently than the smaller finch, G. fortis.

On the other hand, G. fortis is more efficient when it comes to small seeds. Thus, natural selection would favour larger beaked birds in times when caltrops and other large seeded plants are abundant and smaller beaked birds would be favoured when smaller seeded plants are plentiful.

Such a prediction would hold true with respect to both a comparison between these two species, as well as within each species. Thus, not only should G. magnirostris would be favoured over G. fortis when large seeds are abundant, but also larger G. magnirostris would be favoured over smaller G. magnirostris and, similarly, larger G. fortis over smaller G. fortis.

2. Individual Learning

Animals in course of their lifetime learn about everything from food and shelter to predators and familial relationships. Such individual learning represents a second major force and can alter the frequency of Behavior within the lifetime of an organism.

Individual learning can take many forms. Considering the hypothetical case of learning in the context of mating, it has been observed that females of most animals mate with numerous males throughout the course of a lifetime. For example, imagine a female bird mating with various males somehow able to keep track of how many chicks fledged their nest (1, 2, 3 and so on).

Such a female might learn which male is a good mate by keeping track of the number of eggs she laid when associated with each male. As more egg laying is the preference, during later mating opportunities, such female birds would be likely to choose male bird 2 as she has learned that she lays the most eggs after mating with him. In such a case, learning has changed the Behavior of an animal within the course of a life time.

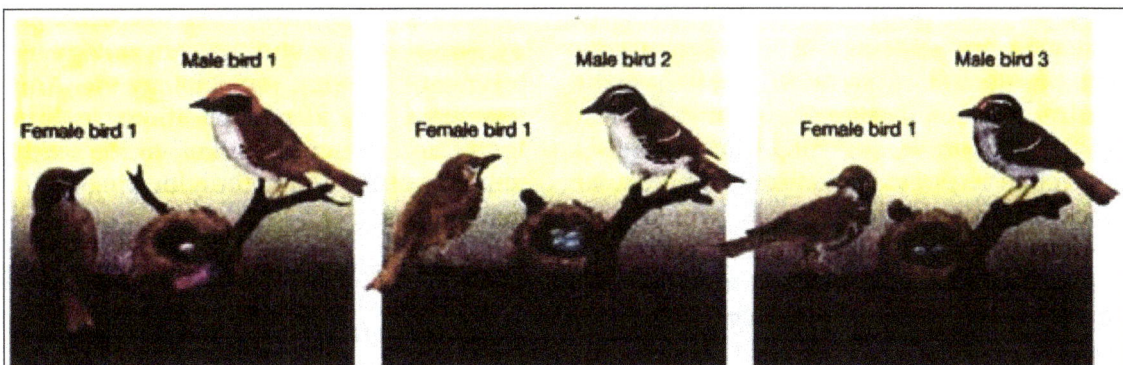

The above is a good example of how learning and natural selection can be intimately tied together. In this example, females change their preference for mates as a result of prior experience and, therefore, learning affected Behavior pattern within a generation.

Such Behavior based on personal experience can not only shift mate choices within a generation, but it can also change the ability to learn, which if genetically coded, can be subjected to natural selection. Natural selection might very well favour the ability to learn about mates, over the lack of such an ability.

3. Cultural Transmission

This is considered to be the third major force affecting animal Behavior, where animals learn something by copying the Behavior of others, through what is known as social learning. Cultural transmission can allow newly acquired traits to spread through populations at a very quick rate, as well as permit the transmission of information across generations rapidly.

The importance of cultural transmission and social learning in animals can be exemplified in the case of foraging in rats. Rats, being scavengers, are often presented with opportunities to sample new food. On one hand, a new food source may be an unexpected rat bounty, while, on the other hand, new foods may be dangerous, because either they contain elements inherently bad for rats, or the rats do not know how a new food should smell.

So the rats face the difficulty to tell whether this new food is fresh or spoiled. To overcome this, foragers often learn critical tidbits about the location and identity of food by interacting with others who have recently returned from a foraging bout.

In case of individual learning, it is certainly possible that if the above Behavior of "copy the diet choice of others" is genetically coded, then the rule might increase in frequency through natural selection. The case of cultural transmission, on the other hand, is more complicated than that of individual learning. The reason being that what an animal learns via individual learning is lost when the animal dies.

The actual information that one learns via individual learning never makes it across generations. However, this is not the case with cultural transmission, where what a single animal does, if copied, can affect individuals many generations later on. Cultural transmission, thus, has both within- and between-generation effects.

Conceptual-theoretical-empirical Approaches in Ethology

Studies in animal behavior tend to use all these three approaches:

In conceptual approaches, ideas generated in different sub-disciplines are imported and combined in a new, cohesive way. Major conceptual advances tend to generate not only new experimental work, but they also reshape the way that a discipline looks at itself.

In many species, like the vervet monkey (Cercopithecus aethiops), exhibits kin selection and mother-offspring bond, where mothers go to extreme lengths to provide for and protect their young offsprings.

This provides a conceptual framework for understanding the special relations that blood relatives share. Individual's total fitness, measured by its genetic contribution to the next generation, is not simply a function of the number of viable offsprings that it produces—rather, it is a combination of the number of young it raises, plus some benefit assigned for any help it provided in raising the offspring of blood relatives.

Theoretical Approach

Theoretical approach to animal Behavior relates to the construction of a mathematical model. A question of interest is: If a list of potential edible items is given, which one should a foraging animal add to its diet and under what conditions? Ethologists have constructed mathematical models of foraging that determine which potential prey items should be taken.

The value assigned to each prey is a composite of energy value (e), handling time (h), and encounter rate (λ) associated with various items to predict which such item should be added to an animal's diet to optimize some quantity, such as energy intake per unit time.

For example, the model predicted that whether or not a low-ranked food item would be added to an animal's diet depended on the availability of the low-ranked item itself. Such as, if for wolf predators, rabbit provided more energy per unit time than did chickens, then the availability of rabbits, not chickens, would determine whether chickens would be added to the wolf's diet.

Empirical studies are designed to test the theories and concepts that have been proposed as explanations for Behavior. Of the many forms the most essential empirical work in ethology is either observational or experimental.

Observational work requires watching and recording of what animals do, but no attempt is made to manipulate or control an ethological or environmental variable. For example, one might go out into a forest and record every action of a particular flock of birds.

In doing so, he would note various Behaviors like foraging, encounter with predators, feeding of nestlings, the time and duration it sits on a particular tree, and so on. From such work, one can visualize-

- The time budget of the birds in study.

- The foraging of the males and females.

- Predict their relation of foraging bouts when predators are present in their vicinity.

- Find the correlation between foraging Behavior and predation pressure, and so on.

However, from the above observational work, it is difficult to speak of what caused what. It may be that there are other variables which might be responsible for the correlation of foraging with predation pressure. To know that, one might have to experimentally manipulate the system.

For example, two areas might be taken. In area 1 the predators may be increased, while in area 2 natural conditions may prevail. Now it is to be seen whether foraging is affected by the increase in predators or not. Therefore, we can then confidently conclude whether increased predation pressure causes decreased foraging activity or not.

NEUROETHOLOGY

Neuroethology is the evolutionary and comparative approach to the study of animal behavior and its underlying mechanistic control by the nervous system. This interdisciplinary branch of behavioral neuroscience endeavors to understand how the central nervous system translates biologically relevant stimuli into natural behavior. For example, many bats are capable of echolocation which is used for prey capture and navigation. The auditory system of bats is often cited as an example for how acoustic properties of sounds can be converted into a sensory map of behaviorally relevant features of sounds. Neuroethologists hope to uncover general principles of the nervous system from the study of animals with exaggerated or specialized behaviors.

As its name implies, neuroethology is a multidisciplinary field composed of neurobiology (the study of the nervous system) and ethology (the study of animal behavior in natural conditions). A central theme of the field of neuroethology, delineating it from other branches of neuroscience, is this focus on natural behavior, which may be thought of as those behaviors generated through means of natural selection (i.e. finding mates, navigation, locomotion, predator avoidance) rather than behaviors in disease states, or behavioral tasks that are particular to the laboratory.

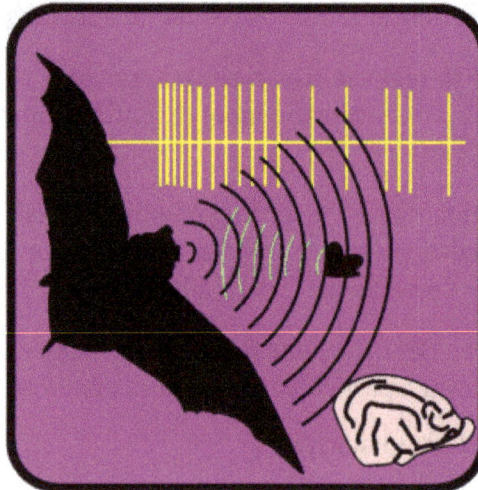

Echolocation in bats is one model system in neuroethology.

Neuroethology is an integrative approach to the study of animal behavior that draws upon several disciplines. Its approach stems from the theory that animals' nervous systems have evolved to address problems of sensing and acting in certain environmental niches and that their nervous systems are best understood in the context of the problems they have evolved to solve. In accordance with Krogh's principle, neuroethologists often study animals that are "specialists" in the behavior the researcher wishes to study e.g. honeybees and social behavior, bat echolocation, owl sound localization, etc.

The scope of neuroethological inquiry might be summarized by Jörg-Peter Ewert, a pioneer of neuroethology, when he considers the types of questions central to neuroethology in his 1980 introductory text to the field:

- How are stimuli detected by an organism?

- How are environmental stimuli in the external world represented in the nervous system?

- How is information about a stimulus acquired, stored and recalled by the nervous system?

- How is a behavioral pattern encoded by neural networks?

- How is behavior coordinated and controlled by the nervous system?

- How can the ontogenetic development of behavior be related to neural mechanisms?

Often central to addressing questions in neuroethology are comparative methodologies, drawing upon knowledge about related organisms' nervous systems, anatomies, life histories, behaviors and environmental niches. While it is not unusual for many types of neurobiology experiments to give rise to behavioral questions, many neuroethologists often begin their research programs by observing a species' behavior in its natural environment. Other approaches to understanding nervous systems include the systems identification approach, popular in engineering. The idea is to stimulate the system using a non-natural stimulus with certain properties. The system's response to the stimulus may be used to analyze the operation of the system. Such an approach is useful for linear systems, but the nervous system is notoriously nonlinear, and neuroethologists argue that such an approach is limited. This argument is supported by experiments in the auditory system, which show that neural responses to complex sounds, like social calls, can not be predicted by the knowledge gained from studying the responses due to pure tones (one of the non-natural stimuli favored by auditory neurophysiologists). This is because of the non-linearity of the system.

Modern neuroethology is largely influenced by the research techniques used. Neural approaches are necessarily very diverse, as is evident through the variety of questions asked, measuring techniques used, relationships explored, and model systems employed. Techniques utilized since 1984 include the use of intracellular dyes, which make maps of identified neurons possible, and the use of brain slices, which bring vertebrate brains into better observation through intracellular electrodes. Currently, other fields toward which neuroethology may be headed include computational neuroscience, molecular genetics, neuroendocrinology and epigenetics. The existing field of neural modeling may also expand into neuroethological terrain, due to its practical uses in robotics. In all this, neuroethologists must use the right level of simplicity to effectively guide research towards accomplishing the goals of neuroethology.

Critics of neuroethology might consider it a branch of neuroscience concerned with 'animal trivia'. Though neuroethological subjects tend not to be traditional neurobiological model systems (i.e. Drosophila, C. elegans, or Danio rerio), neuroethological approaches emphasizing comparative methods have uncovered many concepts central to neuroscience as a whole, such as lateral inhibition, coincidence detection, and sensory maps. The discipline of neuroethology has also discovered and explained the only vertebrate behavior for which the entire neural circuit has been described: the electric fish jamming avoidance response. Beyond its conceptual contributions, neuroethology makes indirect contributions to advancing human health. By understanding simpler nervous systems, many clinicians have used concepts uncovered by neuroethology and other branches of neuroscience to develop treatments for devastating human diseases.

The field of neuroethology owes part of its existence to the establishment of ethology as a unique discipline within the discipline of Zoology. Although animal behavior had been studied since the time of Aristotle, it was not until the early twentieth century that ethology finally became distinguished from natural science (a strictly descriptive field) and ecology. The main catalysts behind this new distinction were the research and writings of Konrad Lorenz and Niko Tinbergen.

Konrad Lorenz was born in Austria in 1903, and is widely known for his contribution of the theory of fixed action patterns (FAPs): endogenous, instinctive behaviors involving a complex sequence of movements that are triggered ("released") by a certain kind of stimulus. This sequence always proceeds to completion, even if the original stimulus is removed. It is also species-specific and performed by nearly all members. Lorenz constructed his famous "hydraulic model" to help illustrate this concept, as well as the concept of action specific energy, or drives.

Niko Tinbergen was born in the Netherlands in 1907 and worked closely with Lorenz in the development of the FAP theory; their studies focused on the egg retrieval response of nesting geese. Tinbergen performed extensive research on the releasing mechanisms of particular FAPs, and used the bill-pecking behavior of baby herring gulls as his model system. This led to the concept of the supernormal stimulus. Tinbergen is also well known for his four questions that he believed ethologists should be asking about any given animal behavior; among these is that of the mechanism of the behavior, on a physiological, neural and molecular level, and this question can be thought of in many regards as the keystone question in neuroethology. Tinbergen also emphasized the need for ethologists and neurophysiologists to work together in their studies, a unity that has become a reality in the field of neuroethology.

Unlike behaviorism, which studied animals' reactions to non-natural stimuli in artificial, laboratory conditions, ethology sought to categorize and analyze the natural behaviors of animals in a field setting. Similarly, neuroethology asks questions about the neural bases of naturally occurring behaviors, and seeks to mimic the natural context as much as possible in the laboratory.

Although the development of ethology as a distinct discipline was crucial to the advent of neuroethology, equally important was the development of a more comprehensive understanding of neuroscience. Contributors to this new understanding were the Spanish Neuroanatomist, Ramon y Cajal, and physiologists Charles Sherrington, Edgar Adrian, Alan Hodgkin, and Andrew Huxley. Charles Sherrington, who was born in Great Britain in 1857, is famous for his work on the nerve synapse as the site of transmission of nerve impulses, and for his work on reflexes in the spinal cord. His research also led him to hypothesize that every muscular activation is coupled to an inhibition of the opposing muscle. He was awarded a Nobel Prize for his work in 1932 along with Lord Edgar Adrian who made the first physiological recordings of neural activity from single nerve fibers.

Alan Hodgkin and Andrew Huxley, are known for their collaborative effort to understand the production of action potentials in the giant axons of squid. The pair also proposed the existence of ion channels to facilitate action potential initiation, and were awarded the Nobel Prize in 1963 for their efforts.

As a result of this pioneering research, many scientists then sought to connect the physiological aspects of the nervous and sensory systems to specific behaviors. These scientists – Karl von Frisch, Erich von Holst, and Theodore Bullock – are frequently referred to as the "fathers" of neuroethology. Neuroethology did not really come into its own, though, until the 1970s and 1980s, when new, sophisticated experimental methods allowed researchers such as Masakazu Konishi, Walter Heiligenberg, Jörg-Peter Ewert, and others to study the neural circuits underlying verifiable behavior.

Modern Neuroethology

The International Society for Neuroethology represents the present discipline of neuroethology, which was founded on the occasion of the NATO-Advanced Study Institute "Advances in Vertebrate Neuroethology" organized by J.-P. Ewert, D.J. Ingle and R.R. Capranica, held at the University of Kassel in Hofgeismar, Germany. Its first president was Theodore H. Bullock. The society has met every three years since its first meeting in Tokyo in 1986.

Its membership draws from many research programs around the world; many of its members are students and faculty members from medical schools and neurobiology departments from various universities. Modern advances in neurophysiology techniques have enabled more exacting approaches in an ever-increasing number of animal systems, as size limitations are being dramatically overcome. Survey of the most recent congress of the ISN meeting symposia topics gives some idea of the field's breadth:

- Comparative aspects of spatial memory (rodents, birds, humans, bats).

- Influences of higher processing centers in active sensing (primates, owls, electric fish, rodents, frogs).

- Animal signaling plasticity over many time scales (electric fish, frogs, birds).

- Song production and learning in passerine birds.

- Primate sociality.

- Optimal function of sensory systems (flies, moths, frogs, fish).

- Neuronal complexity in behavior (insects, computational).

- Contributions of genes to behavior (Drosophila, honeybees, zebrafish).

- Eye and head movement (crustaceans, humans, robots).

- Hormonal actions in brain and behavior (rodents, primates, fish, frogs, and birds).

- Cognition in insects (honeybee).

Application to Technology

Neuroethology can help create advancements in technology through an advanced understanding of animal behavior. Model systems were generalized from the study of simple and related animals to humans. For example, the neuronal cortical space map discovered in bats, a specialized champion of hearing and navigating, elucidated the concept of a computational space map. In addition, the discovery of the space map in the barn owl led to the first neuronal example of the Jeffress model. This understanding is translatable to understanding spatial localization in humans, a mammalian relative of the bat. Today, knowledge learned from neuroethology are being applied in new technologies. For example, Randall Beer and his colleagues used algorithms learned from insect walking behavior to create robots designed to walk on uneven surfaces. Neuroethology and technology contribute to one another bidirectionally.

Neuroethologists seek to understand the neural basis of a behavior as it would occur in an animal's natural environment but the techniques for neurophysiological analysis are lab-based, and cannot be performed in the field setting. This dichotomy between field and lab studies poses a challenge for neuroethology. From the neurophysiology perspective, experiments must be designed for controls and objective rigor, which contrasts with the ethology perspective – that the experiment be applicable to the animal's natural condition, which is uncontrolled, or subject to the dynamics of the environment. An early example of this is when Walter Rudolf Hess developed focal brain stimulation technique to examine a cat's brain controls of vegetative functions in addition to other behaviors. Even though this was a breakthrough in technological abilities and technique, it was not used by many neuroethologists originally because it compromised a cat's natural state, and, therefore, in their minds, devalued the experiments' relevance to real situations.

When intellectual obstacles like this were overcome, it led to a golden age of neuroethology, by focusing on simple and robust forms of behavior, and by applying modern neurobiological methods to explore the entire chain of sensory and neural mechanisms underlying these behaviors. New technology allows neuroethologists to attach electrodes to even very sensitive parts of an animal such as its brain while it interacts with its environment. The founders of neuroethology ushered this understanding and incorporated technology and creative experimental design. Since then even indirect technological advancements such as battery-powered and waterproofed instruments have allowed neuroethologists to mimic natural conditions in the lab while they study behaviors objectively. In addition, the electronics required for amplifying neural signals and for transmitting them over a certain distance have enabled neuroscientists to record from behaving animals performing activities in naturalistic environments. Emerging technologies can complement neuroethology, augmenting the feasibility of this valuable perspective of natural neurophysiology.

Another challenge, and perhaps part of the beauty of neuroethology, is experimental design. The value of neuroethological criteria speak to the reliability of these experiments, because these discoveries represent behavior in the environments in which they evolved. Neuroethologists foresee future advancements through using new technologies and techniques, such as computational neuroscience, neuroendocrinology, and molecular genetics that mimic natural environments.

Computational Neuroethology

Computational neuroethology (CN or CNE) is concerned with the computer modelling of the neural mechanisms underlying animal behaviors. Computational neuroethology was first argued for in depth by Randall Beer and by Dave Cliff both of whom acknowledged the strong influence of Michael Arbib's Rana Computatrix computational model of neural mechanisms for visual guidance in frogs and toads.

CNE systems work within a closed-loop environment; that is, they perceive their (perhaps artificial) environment directly, rather than through human input, as is typical in AI systems. For example, Barlow et al. developed a time-dependent model for the retina of the horseshoe crab Limulus polyphemus on a Connection Machine (Model CM-2). Instead of feeding the model retina with idealized input signals, they exposed the simulation to digitized video sequences made underwater, and compared its response with those of real animals.

INSTINCT BEHAVIOR

Behaviors that are closely controlled by genes with little or no environmental influence are called innate behaviors. These are behaviors that occur naturally in all members of a species whenever they are exposed to a certain stimulus. Innate behaviors do not have to be learned or practiced. They are also called instinctive behaviors. An instinct is the ability of an animal to perform a behavior the first time it is exposed to the proper stimulus. For example, a dog will drool the first time—and every time—it is exposed to food.

Significance of Innate Behavior

Innate behaviors are rigid and predictable. All members of the species perform the behaviors in the same way. Innate behaviors usually involve basic life functions, such as finding food or caring for offspring. Several examples are shown in figure below. If an animal were to perform such important behaviors incorrectly, it would be less likely to survive or reproduce.

These innate behaviors are necessary
for survival or reproduction.

Intelligence and Innate Behavior

Innate behaviors occur in all animals. However, they are less common in species with higher levels of intelligence. Humans are the most intelligent species, and they have very few innate behaviors. The only innate behaviors in humans are reflexes. A reflex is a response that always occurs when a certain stimulus is present. For example, a human infant will grasp an object, such as a finger, that is placed in its palm. The infant has no control over this reaction because it is innate. Other than reflexes such as this, human behaviors are learned–or at least influenced by experience–rather than being innate.

LEARNED BEHAVIOR

Learning is a change in behavior that occurs as a result of experience. Compared with innate behaviors, learned behaviors are less rigid. Many learned behaviors can be modified to suit changing conditions. For example, drivers may have to change how they drive (a learned behavior) when roads are wet or icy, otherwise they may risk losing control of their vehicle. Because learned behaviors can be modified when the environment changes, they are generally more adaptive than innate behaviors. Species that are more intelligent typically have a greater proportion of behaviors that are learned rather than innate.

Types of Learning

Animals may learn behaviors in a variety of ways. Some ways in which animals learn are relatively simple. Others are very complex. Types of learning include the following:

- Habituation.

- Sensitization.

- Classical conditioning.

- Operant conditioning.

- Observational learning.

- Play.

- Insight learning.

Habituation and Sensitization

One of the simplest ways that animals learn is through habituation. In this type of learning, animals decrease the frequency of a behavior in response to a repeated stimulus. This occurs when the behavior does not result in some type of benefit or reward. Habituation has been demonstrated to occur in virtually every species of animal. It is adaptive because responding to a stimulus when there is no benefit or reward is a waste of energy.

Coyotes are becoming increasingly habituated to humans, especially when humans feed them, so they are no longer afraid to approach human neighborhoods. This can be quite dangerous, as coyotes are known to attack livestock, pets, and, sometimes, even humans.

An example of habituation is the behavior of certain species of small songbirds when presented with a stuffed owl or similar "predator." If a stuffed owl is placed in their cage, the birds first respond as though it were a real predator. They act frightened and try to escape. Over time, as the stuffed owl remains in the cage without moving, the birds show less response. They become habituated to the presence of the stuffed owl. A similar example of habituation is when coyotes invade human neighborhoods. They have become habituated to humans in these locations, so they are no longer afraid to approach.

Sensitization is the opposite of habituation. In sensitization, an animal learns to react more often or more strongly to a repeated stimulus. For example, exposure to painfully loud sounds causes an animal to respond strongly. The animal may act agitated and try to escape from the source of the sounds. If the loud sounds are followed by lesser sounds that are not painful, the animal may respond to them just as strongly. If this occurs, the animal has become sensitized to sounds.

Scientists have demonstrated that sensitization occurs because of changes in nerve cells and nerve pathways. These changes take place after nerves have been stimulated repeatedly. This sometimes occurs in a person who has had a painful injury. With repeated stimulation of the nerves, sensitization occurs, and the pain continues even after the injury is healed.

Classical Conditioning

Classical conditioning is a type of learning in which an animal learns to associate one stimulus with another. In this type of learning, a stimulus that normally produces a particular behavior is linked with a second stimulus. The second stimulus is something neutral to which the animal does not normally respond. If the animal is repeatedly exposed to both stimuli together, it learns to associate the two stimuli. Because of this association, the animal will respond to the second stimulus alone in the same way that it responds to the normal stimulus.

A well-known example of classical conditioning is the work of the Russian scientist Ivan Pavlov. In the late 1800s and early 1900s, Pavlov investigated behavior in dogs. As mentioned above, dogs instinctively drool when they see or smell food. They also drool when they think that they are about to be fed. Pavlov conditioned dogs to drool when they heard a particular sound, such as a bell or whistle. He made the sound just before he fed the dogs, and the dogs learned to associate the sound with food. In a short time, they started drooling as soon as they heard the sound. Because of Pavlov's research in this area, classical conditioning is sometimes called Pavlovian conditioning.

Another example of classical conditioning is called conditioned taste aversion. Animals may learn not to eat certain foods if they have ever become ill after eating them. This is an adaptive trait because it may help them avoid foods that are poisonous. For example, animals that vomit after eating a particular type of berry may learn to avoid eating berries of this type in the future. They become conditioned to avoid the berries because they have learned to associate the berries with vomiting.

Operant Conditioning

In operant conditioning, an animal learns either to perform a behavior that is rewarded or to stop performing a behavior that is punished. One of the first scientists to investigate this type of learning was Edward Thorndike. In the early 1900s, Thorndike investigated learning in cats. He placed the cats in "puzzle boxes" that he had constructed. It was difficult for the cats to find their way out of the mazelike boxes, but they kept trying because they did not like being confined. When first placed in one of the boxes, a cat needed a long time to find the way out. However, the cat needed less and less time with repeated trials. Through trial and error, the cat learned how to escape from the box.

Beginning in the 1930s, another scientist, named B.F. Skinner, did similar research with rats. He placed rats in a box containing a bar. If the rats stepped on the bar, a pellet of food was released. When first placed in the box, a rat did not know that stepping on the bar would release a food pellet. Sooner or later, the rat would accidentally step on the bar and be rewarded with food. Before long, the rat learned that a food pellet would be released each time it stepped on the bar. After that, it stepped on the bar repeatedly in order to get the food.

In operant conditioning, the reward may be something positive that is gained (food in the case of Skinner's rats) or something negative that is avoided (confinement in the case of Thorndike's cats). In either case, a behavior is learned through trial and error because it is reinforced by a reward. Operant conditioning can also use punishment to discourage a behavior. A punishment is something unpleasant or painful. An example of this occurs when cows are placed in a pasture surrounded by an electrified fence. The fence alone is inadequate to keep them in the pasture. It is just a single strand of wire strung between posts that are several feet apart. However, when the cows touch the fence, they receive an electric shock. They soon learn from the punishment (the shock) to stay away from the fence (the behavior).

Can you think of a behavior that you learned from operant conditioning? May be you learned that taking notes in class results in better grades on exams. Getting better grades (the reward) may have reinforced note taking (the behavior). You probably also learned to avoid behaviors that were punished. For example, when you were younger, fighting with a sibling may have resulted in a time-out. After repeated time-outs (the punishment), you may have learned to avoid fighting with your sibling (the behavior). Did you ever touch your tongue to a metal object on a cold winter day? If you did, then you know that your tongue "sticks" to the metal and that pulling your tongue away from the metal is very painful. If you learned to avoid this behavior because of the pain, then this is also an example of operant conditioning.

Observational Learning

Perhaps you learned to avoid touching your tongue to a freezing metal object because you saw another child do it and realized how painful it was from the other child's reaction. If so, then you learned not to do it yourself through observational learning. This type of learning involves observing the behavior of another individual and either copying the behavior or avoiding it. Most studies of observational learning have focused on behaviors that are copied. Canadian psychologist Albert Bandura is world renowned for his investigations of observational learning in humans. According to Bandura, learning a behavior by observing it in someone else requires four conditions to be met:

- An individual (the observer) must pay attention to the behavior of another individual (the model).

- The observer must be able to remember what the model has done.

- The observer must have the ability or skills to perform the behavior.

- The observer must be motivated and have the opportunity to perform the behavior.

Because of these conditions, observational learning requires considerable intelligence and is found most often in humans. However, observational learning has also been observed in many other species of animal. For example, wolves and other predatory animals that hunt in packs learn hunting skills through observational learning. Young animals observe and copy the behavior of older animals when they hunt together.

Another example of observational learning involves Japanese macaques. In the 1960s, a group of researchers started placing sweet potatoes on a sandy beach to lure macaques out of a nearby forest. Soon the macaques started coming out on the beach to eat the sweet potatoes. At first, the macaques just brushed the sand off the sweet potatoes before eating them. Then, about a year later, a female macaque was observed washing sweet potatoes in the ocean before eating them. Her behavior was observed and copied by other macaques in the troop. Before long, all the macaques in the troop were washing their sweet potatoes in the ocean. When these monkeys gave birth, their offspring also learned this behavior by observing and copying the behavior of their parents and other adults.

Japanese macaques have learned to wash sweet potatoes in the ocean before eating them. This behavior was first noted in one macaque. Other macaques in the troop soon learned the behavior through observational learning.

Many human behaviors are learned by children through observation and mimicking the behaviors of the people around them. What behaviors have you learned in this way? For example, as a young child, did you learn how to tie your shoes by watching your parents or older siblings tie their shoes? Did you learn how to solve a math problem by watching your teacher solve one like it? Did you learn how to play a video game by observing a friend play the game? If so, you were learning the behaviors through observational learning.

Play

Playing a video game is just one of the many ways in which humans may play. Play involves behaviors that have no particular goal except enjoyment or satisfaction. Play is not restricted to humans. Most mammals and many birds also play when they are young. The drive to play seems to be innate in many species. You have probably seen kittens, like the one in figure below, playing with a toy. Play may also involve other animals. For example, kittens often play with their littermates. Like other predatory animals — including lions and some species of bears — kittens chase, pounce on, and wrestle with one another. Prey animals such as deer and zebras play somewhat differently. They run, leap, and kick their hind legs when they play.

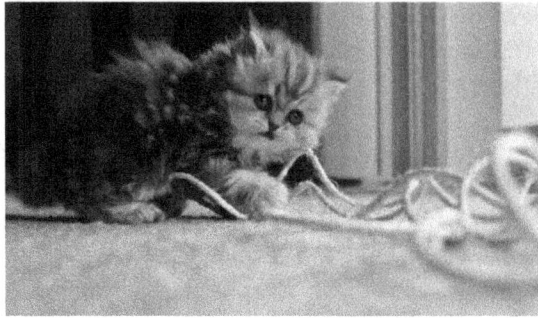

Kittens often play with toys.

Play is one way that young animals develop the skills needed during adulthood. How does playing with a toy prepare this kitten for catching mice as an adult?

This adult cat is trying to catch a mouse.

Play has risks and costs. It may increase exposure to predators and lead to injury. It also requires energy. For play to have evolved, it is likely to have significant benefits that outweigh these drawbacks. How could play be beneficial? Play is actually a form of learning. Through play, young animals learn important skills. By chasing, pouncing on, and wrestling with one another, kittens and other young predators are learning how to catch prey. By running, leaping, and kicking their hind legs, young deer, zebras, and other prey animals are learning how to escape or ward off attacks from predators. Play can also be beneficial by helping young animals develop muscles and improve their physical fitness.

Insight Learning

Several species of animals have been observed using insight learning. Insight learning is the use of past experiences and reasoning to solve problems. Unlike operant conditioning, insight learning does not involve trial and error. Instead, an animal thinks through a solution to a problem based on previous experience. The solution often comes in a flash of insight. Insight learning requires relatively great intelligence. Species most likely to learn in this way include species of apes (chimpanzees, gorillas, and orangutans), crows, and humans.

Examples of insight learning in apes in the wild include chimpanzees "fishing" for termites. The animals place a "tool" such as a twig in a hole of termite mound, and the termites crawl onto the twig. Then the chimpanzee withdraws the twig from the hole and eats the termites. Chimpanzees may also use leaves to sop up drinking water and rocks to smash nuts. They have even been observed sharpening sticks with their teeth and using them as spears to kill small animals for food. Another example of tool use in chimpanzees in the wild is shown in figure below. Gorillas may use tools to solve problems as well. For example, they have been observed using small branches as walking sticks and as measuring sticks to test the depth of water before wading through it. Orangutans use small sticks to get at edible seeds inside prickly fruits without being pricked. They may also use leaves to make rain-hats and roofs over their sleeping nests.

This chimpanzee is using a stick as a tool to obtain food that is out of reach. This is an example of insight learning.

Crows and related species of birds have large brains relative to their body size and are noted for their intelligence. They also use insight learning to solve problems. For example, a crow was observed placing nuts on the crosswalk of a busy street, where cars would drive over them and crack the shells. After the traffic light turned red, the crow entered the street to retrieve the nutmeats. Crows have also been seen working together to eat food scraps in a trashcan. Some of the crows propped up the lid of the can, while the others ate the scraps inside. Then the crows switched places so that all of them had a chance to eat.

Homo sapiens have larger brains (for their body size) and are more intelligent than any other species. Humans are also known for their incredible ability to use insight learning to solve problems, ranging from learning how to start a fire to putting a man on the moon. Think about problems you have solved in the past. Perhaps you learned how to use a new computer program through reasoning and past experience with other computer programs. You may also have used reasoning and previous experience to solve math problems or attain the next level of a video game.

Behavior Influenced by both Genes and Learning

The ability to learn behaviors in all these different ways depends on intelligence, which is at least partly determined by genes. However, the environment also plays a role in learned behaviors. For example, classical conditioning depends on the presence of a suitable reward or punishment, and observational learning depends on the behavior of potential models.

A type of behavior that clearly shows the influence of both genes and environment is imprinting. Like instinct, imprinting results in fixed, lifelong behaviors after exposure to a stimulus. Also like instinct, imprinting leads to full-blown behaviors after the first exposure to the stimulus. However, the environment also plays a role in imprinting. The animal must be exposed to the proper stimulus during a period of development called the critical period. This period is commonly a few days or weeks in early life. In addition, the type of stimulus that triggers an imprinted behavior determines how the behavior is performed. Therefore, imprinting depends on both instinct and learning and may vary with the environment.

Imprinting often occurs in species in which the young are fairly mobile early in life. Animals in which imprinting occurs include species of aquatic birds and herbivorous mammals. In these species, baby animals instinctively follow and become attached to whatever large moving object they first see during the critical period. The moving object is usually their mother. Becoming imprinted on the mother helps ensure their survival. They are unlikely to wander off after they have become imprinted on her. However, the infants may become imprinted on any other large moving object if they see it first during the critical period. This could be a balloon blowing across the ground or a human walking nearby. Once attachment to the object occurs through imprinting, the behavior is fixed and cannot be changed. The animal remains attached to the object for life. Moose calves also imprint on their mothers. If a moose calf were to imprint on a human, it would continue its attachment to humans rather than members of its own species. As an adult, the moose would be likely to try to attract humans, rather than other moose, as mates.

If a human were the first large moving object a moose calf saw during a critical
period in its development, the moose calf would imprint on the human.

There are many other examples of imprinting in animals:

- Puppies must be exposed to humans within the first two or three months of life, or they will never become socialized and be suitable as human companions.

- In birds called zebra finches, chicks raised with another species of bird, rather than with other zebra finches, will imprint on birds of the other species. As adults, the finches will try to mate with birds of the other species instead of with other zebra finches.

From all of these examples, it appears that imprinting may or may not be adaptive. Whether it is adaptive depends on the timing and type of stimulus that triggers the behavior. If the stimulus occurs during the critical period and is appropriate, then imprinting is likely to be adaptive and increase the fitness of the imprinting animal. If the stimulus occurs outside of the critical period or is inappropriate, then imprinting may not be adaptive and may reduce fitness.

ANIMAL SOCIAL BEHAVIOR

Animal social Behavior is the suite of interactions that occur between two or more individual animals, usually of the same species, when they form simple aggregations, cooperate in sexual or parental Behavior, engage in disputes over territory and access to mates, or simply communicate across space.

Social Behavior is defined by interaction, not by how organisms are distributed in space. Clumping of individuals is not a requirement for social Behavior, although it does increase opportunities for interaction. When a lone female moth emits a bouquet of pheromones to attract male potential mates, she is engaging in social Behavior. When a male red deer (Cervus elaphus) gives a loud roar to signal dominance and keep other males away, he is also being social.

Animal social Behavior has piqued the interest of animal behaviorists and evolutionary biologists, and it has also engaged the public, thanks to life science filmmakers who captured the drama and stunning diversity of animal social interactions in documentaries and other media programs.

General Characteristics

Social Behavior ranges from simple attraction between individuals to life in complex societies characterized by division of labour, cooperation, altruism, and a great many individuals aiding the reproduction of a relative few. The most widely recognized forms of social Behavior, however, involve interaction within aggregations or groups of individuals. Social Behaviors, their adaptive value, and their underlying mechanisms are of primary interest to scientists in the fields of animal Behavior, behavioral ecology, evolutionary psychology, and biological anthropology.

Army ants (Eciton).

The word social often connotes amicable interaction, accounting for the common misconception that social Behavior always involves cooperation toward some mutually beneficial end. Biologists no longer believe that cooperative Behaviors necessarily evolve for the good of the species. Instead, they believe that the unit of natural selection is usually the individual and that social Behavior is fraught with competition. English naturalist Charles Darwin, who first brought evolution by natural selection to the attention of the world, introduced this paradigm for thinking about social Behavior, noting that it is the best competitors within a species, the "fittest" individuals, that survive and reproduce. Once genetics was integrated into this concept of evolution, it became apparent that such individuals will transmit the most copies of their genes to future generations.

Consistent with Darwin's ideas, social organisms are often seen to be fiercely competitive and aggressive. For example, friendly interactions among children on a playground can quickly dissolve into fierce competition if there are too few balls or swings. In addition, intense competitive interactions resulting in bodily harm can occur even among family members. Social Behavior is designed to enhance an individual's ability to garner resources and form the alliances that help it to survive and to reproduce. The modern view of social Behavior is that it is a product of the competing interests of the individuals involved. Individuals evolve the capacity to behave selfishly and to cooperate or compete when it benefits them to do so. A delicate balance of cooperative and competitive Behaviors is thus expected to characterize animal societies.

Categorizing the Diversity of Social Behavior

Social Behavior encompasses a wide variety of interactions, from temporary feeding aggregations or mating swarms to multigenerational family groups with cooperative brood care. Over the years, there have been many attempts to classify the diversity of social interactions and understand the evolutionary progression of social Behavior.

A small group of European bison (Bison bonasus) grazing near the mountains.

A series of veteran American entomologists—starting in the 1920s with William Morton Wheeler and continuing into the 1970s with Howard Evans, Charles Michener, and E.O. Wilson—developed a categorization of sociality following two routes, called the parasocial sequence and the subsocial sequence. This classification is based primarily on the involvement of insect parents with their young, whereas classifications of vertebrate sociality are frequently based on spacing Behavior or mating system. Both routes culminate in "eusociality," a system in which the young are cared for cooperatively and the society is segregated into different castes that provide different services.

In the parasocial sequence, adults of the same generation assist one another to varying degrees. At one end of the spectrum are females of communal species; these females cooperate in nest construction but rear their broods separately. In quasisocial species, broods are attended cooperatively, and each female may still reproduce. Semisocial species also practice cooperative brood care, but they possess within the colony a worker caste of individuals that never reproduce. Eusocial species typically engage in cooperative brood care; in addition, they have distinct castes that perform different functions and an overlap of generations within the colony.

The subsocial sequence, the alternate route to eusociality, involves increasingly close association between females and their offspring. In primitively subsocial species, the female provides direct care for a time but departs before the young emerge as adults. This stage is followed by two intermediate subsocial stages: one where the care of young is extended to the point where the mother is present when her offspring mature, and the other where offspring are retained that assist in the rearing of additional broods. At the eusocial end of this sequence, some mature offspring are differentiated into a permanently sterile worker caste—a stage that mirrors the same eusocial outcome achieved by the parasocial sequence described above.

E.O. Wilson, whose Sociobiology: The New Synthesis provided a blueprint for research in this field when it was published in 1975, felt that general classifications of societies invariably fail because they depend on the qualities chosen to divide species, which vary markedly from group to group. Instead, Wilson compiled a set of 10 essential qualities of sociality, including (1) group size, (2) distributions of different age and sex classes, (3) cohesiveness, (4) amount and pattern of connectedness, (5) "permeability," or the degree to which societies interact with one another, (6) "compartmentalization," or the extent to which subgroups operate as discrete units, (7) differentiation of roles among group members, (8) integration of Behaviors within groups, (9) communication and information flow, and (10) fraction of time devoted to social Behavior as opposed to individual maintenance. These overlapping qualities of societies provide a good indication of the complexities involved with classifying, much less understanding, the highly varied social Behavior of animals.

While categories of social Behavior can be useful, they can also be confusing and misleading. The current tendency is to view sociality as a multifaceted continuum from simple aggregations to the highly organized and complex levels of social organization found in eusocial species. Biologists interested in sociality focus on how cooperation increases an individual's genetic legacy, either by increasing its ability to produce offspring directly or by increasing the number of offspring produced by relatives.

The Range of Social Behavior in Animals

The range of social Behavior is best understood by considering how sociality benefits the individuals involved. Because interacting with other individuals is inherently dangerous and potentially costly, both the costs and benefits of social Behavior and the costs and benefits of aggregating with others play a role in the evolution of aggregation.

On the positive side, aggregation may provide individuals with increased access to food through information sharing and cooperative defense against non-group members. Conversely, close contact with members of the same species increases the risk of cannibalism, parasitism, and disease. This is illustrated by studies of cliff swallows (Hirundo pyrrhonota), which suggest that the original benefit of nesting near other individuals and forming colonies was information sharing and increased ability to exploit a highly variable insect food resource. Once colonies were formed, other benefits arose including more efficient detection of predators. These benefits are countered by several costs of coloniality, including increased susceptibility to ectoparasites (that is, parasites such as fleas and ticks that live on the body surface of the host), increased incidence of food stealing (kleptoparasitism), and the need to travel greater distances to foraging areas.

Some costs and benefits overlap. For example, coloniality increases the opportunity for some males to mate with females other than their primary mate (extra-pair matings); it is a benefit for the males succeeding in obtaining extra-pair matings and a cost to the cuckolded males. Similarly, coloniality allows some females to lay eggs in the nests of other females in the colony (conspecific brood parasitism); it is beneficial to them but costly to the parasitized pair. The outcome of these multiple factors, which include the extent to which each individual involved is affected, is a delicate balance leading to wide variation in group sizes ranging from solitary nesting to nesting in colonies of several thousand pairs.

In groups, potential benefits of efficient predator detection and prey acquisition may be diluted by the costs of sharing food and reproductive opportunities. Furthermore, all individuals within groups are not equal; as dominant individuals monopolize a group's resources more effectively, group living becomes less beneficial for subordinates. If social Behavior is to be maintained in a population, even subordinate individuals must gain more from being social than from leaving the group and trying to survive and reproduce on their own.

On the other hand, aggregation may be advantageous due to the energy saved by huddling during cold weather, increased survival through group defense, or increased ability to acquire, hold, and make efficient use of resources. Animals may aggregate by mutual attraction to each other, by mutual attraction to limited resources, or as a side effect of having hatched from eggs laid together in a clutch. In some cases more than one mechanism of attraction is involved. For example, bark beetles (family Scolytidae) form large aggregations by mutual attraction to the bark of a fallen log and also to the odours of other members of their species.

Regardless of the mechanism of attraction, once animals are grouped there is selection to evolve increasing degrees of communication, cooperation, individual recognition, and efficiency to better exploit the potential advantages of group living. For example, in some treehopper (family Membracidae) aggregations, nymphs communicate the threat of a predator by using vibrations, which humans can detect only with electronic instruments. A more sophisticated form of communication is found in eastern tent caterpillars (Malacosoma americanum), which rest in a communal tent that increases in size as they grow and add silk. Colony members leave the tent on brief forays to feed on foliage within their tree, at which point they lay chemical trails that other group members follow to locate high-quality feeding sites. A similar kind of central-place foraging is practiced by some colonial birds, such as cliff swallows, among which unsuccessful individuals often cue in on other birds returning to their nest with food and follow them to productive foraging sites.

In addition to feeding and defensive aggregations, some aggregations are based exclusively on mating. These include the explosive breeding assemblages of frogs and toads, the aggregations of male birds and mammals at leks (display sites used only for mating), and various insect aggregations including bees and wasps (order Hymenoptera), flies (order Diptera), and butterflies (superfamily Papilionoidea). Some species in each of these groups congregate at conspicuous landmarks visited by females. Frequently, the aggregation of one sex provides opportunities for the other. For example, when females aggregate due to the clumping of food or nest sites, males are likely to aggregate at these sites as well because they are the most efficient places to find females with which to mate.

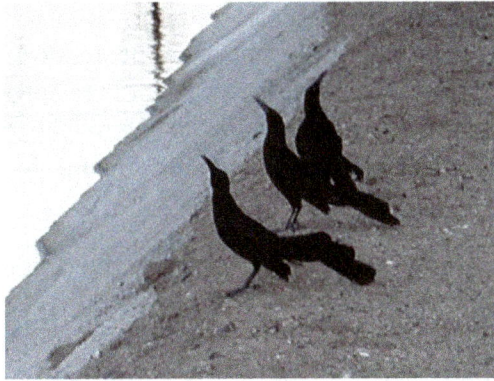

Lek Behavior in three great-tailed grackles (Cassidix mexicanus).

Other groups include flocks or herds that form during migration and coalitions that form due to group advantages in holding or acquiring a reproductive vacancy. Coalitions of male African lions (Panthera leo) that compete for control of groups of females (called prides) are a classic example of the latter. Migration in herds is common and can involve tremendous numbers of individuals. For example, more than one million blue wildebeest (gnu; Connochaetes taurinus) typically migrate in a clockwise fashion over the plains of East Africa, covering a distance of over 2,500 km (about 1,550 miles) each year in search of rain-ripened grass. The record size for migratory aggregations is probably the African desert locust (Schistocerca gregaria), which forms huge swarms covering as much as 200 square km (about 80 square miles); each swarm contains upward of 10 billion individuals moving more or less cohesively in search of food. Among vertebrates, the largest known aggregations were probably those of the now-extinct passenger pigeon (Ectopistes migratorius) of North America. Vast groups of these birds migrating together in search of food, particularly large acorn crops, reportedly exceeded three to five billion individuals.

When group members are genetically related, interactions will potentially involve nepotism, the tendency for individuals to favour kin. Examples of this sort of favouritism include parents favouring their own offspring, siblings forming alliances, and a tendency for individuals to favour their closest relatives. In Siberian jays (Perisoreus infaustus), parents tolerate their offspring on their territories for up to three years, allowing them preferential access to food. Like the African lions mentioned previously, acorn woodpeckers (Melanerpes formicivorus) in southwestern North America and Central America form same-sex sibling coalitions to better compete for reproductive vacancies. European long-tailed tits (Aegithalos caudatus) return home to help feed young still residing in their parents' nests when their own breeding attempts fail. All these Behaviors are facilitated by the genetic relatedness between the individuals involved.

The pinnacle of social Behavior is found in eusocial species. Eusocial species live in multigenerational family groups in which the vast majority of individuals cooperate to aid relatively few reproductive group members (or even a single member). Eusocial Behavior is found in ants, bees, some wasps in the family Vespidae, termites (order Isoptera; sometimes placed in the cockroach order, Blattodea), some thrips (order Thysanoptera), aphids (family Aphididae), and possibly some species of beetles (order Coleoptera). Blesmols, such as the naked mole rat (Heterocephalus glaber) and the Damaraland mole rat (Cryptomys damarensis), engage in truly eusocial Behavior; they are the only vertebrate species known to do so. Eusocial species often exhibit extreme task specialization, which makes colonies potentially very efficient in gathering resources. Workers in eusocial colonies are thought to forgo reproduction due to constraints on independent breeding. Such constraints include shortages of food, territories, protection, skill, nest sites, appropriate weather for breeding, and available mates. Workers may never reproduce during their entire lives, but they nonetheless gain inclusive fitness benefits by aiding the reproduction of a queen, who is typically their mother. Such assistance often takes the form of foraging for food, caring for the young, and maintaining and protecting the nest.

The cost-benefit approach can also be applied to reproductive interactions. For example, the reproductive Behavior of many species is designed to achieve multiple mating, but this Behavior may be associated with certain costs, such as the increased risk of injury or of contracting a sexually transmitted disease. Among many species, mating is essentially promiscuous, with individuals either shedding their gametes into the environment without mating or with individuals pairing briefly, just long enough to effect fertilization. The other extreme includes long-lived animals such as Bewick's swans (Cygnus columbianus), which may live for up to 25 years and mate for life. In addition, it has been shown that within many species some individuals engage in same-sex mating; the individual and group costs and benefits of these Behaviors likewise vary among species. Mating Behavior in animals includes the signaling of intent to mate, the attraction of mates, courtship, copulation, postcopulatory Behaviors that protect a male's paternity, and parental Behavior. Parental Behavior ranges from none to vigilant care by both parents and even by additional group members. Biologists refer to the investment in interactions that influence the likelihood of parenting offspring as "mating effort" and the investment that increases the survival or condition of offspring as "parental effort."

Social Behavior is also involved in social dominance and the maintenance of territories, regardless of whether dominance status or territories are held by individuals or by groups of individuals. Territorial species tend to be distributed over the landscape in a more regular fashion than would be predicted if they used the landscape randomly. An important concept in understanding territorial Behavior is the notion of economic defendability. Economic defendability postulates that, in order to be territorial, the benefits of maintaining exclusive access to a space must outweigh the individual's or group's costs of defending the space from other members of its own kind. In the territorial systems of many species, overt defense in the form of direct aggressive Behavior against intruders has given way to indirect defense in the form of vocalization and scent marking.

Social Behavior is a complex combination of the costs and benefits of living in groups, dominance interactions, conflict between the sexes, nepotism when groups are composed of relatives, and cooperation. The diversity of social Behavior has provided significant material for evolutionary biologists interested in understanding natural selection and the process of evolution.

Social Behavior

Proximate versus Ultimate Causation

Social Behavior is best understood by differentiating its proximate cause (that is, how the Behavior arises in animals) from its ultimate cause (that is, the evolutionary history and functional utility of the Behavior). Proximate causes include hereditary, developmental, structural, cognitive, psychological, and physiological aspects of Behavior. In other words, proximate causes are the mechanisms directly underlying the Behavior. For example, an animal separated from the herd may exhibit Behaviors associated with fear reactions (such as elevated heart rate, shaking, and hypersensitivity to sounds), which cause it to behave in ways that increase its chances of reuniting with the group. The underlying hormonal response, which is triggered by separation from the herd, is a proximate cause of these fear-based Behaviors. In contrast, the ultimate causes of social Behaviors include their evolutionary or historical origins and the selective processes that have shaped their past and current functions. In the case of the isolated herd animal, the development of a better defense against predators that results in increased survival of individuals remaining in groups would be an ultimate cause for the tendency to reunite with the herd.

Dutch-born British zoologist and ethologist Nikolaas Tinbergen was first to clarify these levels of explanation, naming four which he referred to as "survival value," "causation," "development," and "evolutionary history." Tinbergen also emphasized the importance of addressing questions at the appropriate level of explanation. For example, determining the underlying mechanism (causation) of a Behavior does not address the hypotheses regarding its historical origin (evolutionary history) or current survival value. This still causes confusion among evolutionary biologists interested in adaptation, and many examples of unproductive arguments across levels of explanation can be found in the scientific literature.

Strong Inference and the Scientific Study of Social Behavior

Use of the scientific method to study social Behavior permits biologists to deduce the proximate and ultimate functions by using strong inference based on a set of critical predictions. If experiments to test these predictions indicate that the predictions are not met, then the hypothesis is falsified and discarded. If the predictions are met, the hypothesis is supported, but that does not prove it is true.

This is illustrated by examining a question: Why do male birds sometimes adopt and feed offspring of widowed females? One possible explanation is that they have mated with the female and have genetic offspring in the female's nest (current benefits hypothesis). An alternative hypothesis is that the adoptive male gains future benefits because his foster-parenting increases the likelihood that the female will mate with him during her next breeding attempt (future benefits hypothesis). The current benefits hypothesis predicts that some of the female's nestlings were sired by the adoptive father, whereas the future benefits hypothesis predicts that the adoptive male will mate sooner, usually with the widowed female, and produce more offspring in the future than an unpaired male that fails to adopt. While mutually exclusive hypotheses are ideal, in many cases Behaviors have more than one current function and, as in the example of adoption, one or both hypotheses may be true.

Strong inference relies on critical predictions that are capable of distinguishing between alternative hypotheses, whether proximate or ultimate. It also relies on devising clear tests in which each alternative can be falsified by using one or more predictions. In general, predictions can be tested either with data collected from field observations or with experiments. Experiments are considered preferable to field observations because confounding factors are more easily controlled. Unfortunately, manipulations involved in experiments may alter other factors beyond those which the scientist intended, especially where social Behavior is concerned. In order to minimize such problems, researchers take great pains to avoid biases in their experimental procedures and to test their hypotheses by using multiple lines of evidence.

For example, consider the question of why offspring of some species of birds and mammals delay dispersal and remain on their natal territory where they may help raise younger siblings. One of the many basic questions raised by such "helpers-at-the-nest" is the importance of genetic related-ness and kinship to the evolution of the Behavior. Experimentally, cross-fostering young so as to eliminate any genetic relatedness between nestlings and helpers does not typically alter or reduce helping Behavior, but does this demonstrate that kinship is not important? The current thinking on this matter is that cross-fostering leads to a situation where totally unrelated young occur in the nest, a situation that has never been found in the wild. Other studies, meanwhile, have shown that the vast majority of helpers normally feed closely related young. When given the choice, helpers whose own nests have failed preferentially choose to aid closely related young over more distantly related or unrelated young. This Behavior was demonstrated even when the latter were closer to the helper's own failed nesting site. Such results indicate that kin selection plays a key role in the evolution of helping Behavior, despite the experiments suggesting otherwise.

Ultimate Causes of Social Behavior

The advantages of Behaviors such as mating and caring for offspring are obvious in that they in-crease the number and survival of an individual's own young. In contrast, social Behaviors such as living in groups and helping others do not always bear obvious links to individual fitness. Because such Behaviors are complex and paradoxical, their ultimate cause has been a key focus of biolo-gists interested in how social Behavior evolves.

Social interactions can be characterized as mutualism (both individuals benefit), altruism (the altruist makes a sacrifice and the recipient benefits), selfishness (the actor benefits at the expense of the recipient), and spite (the actor hurts the recipient and both pay a cost). Mutualistic asso-ciations pose no serious evolutionary difficulty since both individuals derive benefits that exceed what they would achieve on their own. In general, altruism is less likely to evolve, since a gene for altruism should be selected against. Often individuals acting altruistically are close relatives, in which case the likely resolution of this paradox is kin selection, with altruistic individuals gaining indirect fitness benefits by helping relatives produce additional offspring. Altruism between unre-lated individuals is rare, but it occurs and remains the focus of considerable research. Game theory is often applied to research involving cases of altruism between unrelated individuals.

Reciprocal altruism or reciprocity is one solution to the evolutionary paradox of one individual making sacrifices for another unrelated individual. If individuals interact repeatedly, altruism can be favoured as long as the altruist receives a reciprocal benefit that is greater than its initial cost. Reciprocal altruism can be a potent evolutionary force, but only if there is a mechanism to punish cheaters that accept help without reciprocating. Models of reciprocal altruism suggest that even subtle cheating that is difficult to detect eventually results in the loss of the altruistic trait. Conse-quently, it is not surprising that unambiguous examples of reciprocal altruism outside of humans are rare. Studies have suggested, however, that it plays an important role in the evolution of food sharing by vampire bats (Desmodus rotundus) and the interactions between cleaner fish (Labroi-des dimidiatus) with the client fish they attend. The possibility remains that reciprocity could turn out to be more common than currently recognized.

A second solution for how altruism can evolve among unrelated individuals comes from a study in humans. In this study, individuals punished unrelated cheaters (altruistic punishment), even though they received no material benefit for doing so and were unlikely to interact with them in the future. Furthermore, there may be benefits of advertising one's altruism that allow it to flourish among unrelated individuals. This is suggested by the finding that people are more likely to give blood when they receive a badge advertising their donation. Indirect reciprocation has been used to describe situations in which individuals that give tend to be repaid by individuals other than those they help. This special form of reciprocation can also maintain altruism through the impact of an individual's reputation on his or her likelihood of receiving aid or cooperation in the future. Models indicating the role of reputation in sustaining altruism have been proposed as solutions to the "tragedy of the commons," a key explanation for why gaining the cooperation needed to protect and sustain public resources (such as biological diversity, air and water, and the ozone layer) is so difficult.

Selfish Behavior occurs when one individual benefits at the expense of another. Examples, unsurprisingly, are common. In birds, females sometimes exhibit egg-dumping Behavior or intraspecific brood parasitism (that is, the laying of eggs in nests of other pairs, thus parasitizing their parental care). Even though female birds usually cannot tell their eggs from those of other conspecific females, this sort of parasitism is not particularly common, probably because territoriality and nest guarding help to minimize it. Conspecific brood parasitism, however, occurs in over 30 species of ducks and geese as well as in the northern bobwhite quail (Colinus virginianus), ring-necked pheasant (Phasianus colchicus), wood pigeon (Columba palumbus), European starling (Sturnus vulgaris), cuckoo (Cuculidae), and a variety of other species. Heterospecific brood parasitism is even more common with cuckoos and cowbirds (Molothrus), which lay eggs in the nests of a diversity of other species.

Spite as a social interaction presents an interesting puzzle. It is a Behavior that causes harm to the actor and recipient. Spite is thought to evolve in situations where it serves as a signal of status that helps the actor in the future; in the absence of such future benefits, it should not evolve.

Social Interactions Involving Sex

Mating Behavior describes the social interactions involved in joining gametes (that is, eggs and sperm) in the process of fertilization. In most marine organisms, planktonic gametes are shed (or broadcast) into the sea where they float on the tides and have a small but finite chance of encountering one another. In contrast, the majority of terrestrial animals mate in order to bring together their gametes. On land there has been an evolutionary progression. The earliest land animals needed to return to the water in order to breed. This requirement, however, gave way to the practice of placing sperm packets in the terrestrial environment in locations where they would be picked up by females. While both methods are still used by some species, reproduction in many land animals now involves copulation with internal fertilization. Selectivity on the part of females in externally fertilizing species favours males that engage in Behaviors, such as courtship, which entice females to pick up their sperm. Away from water, the requirement for internal fertilization favours copulation, because it allows males to place their sperm closer to the site of fertilization. The ultimate example of this is traumatic insemination found in bedbugs (family Cimicidae), where males pierce the female's body cavity with their genitalia, placing sperm inside her abdomen. Traumatic insemination is costly for females, with multiple inseminations reducing the female's survival and reproductive success. This indicates that males evolved this strategy at the female's expense, resulting in a persistent conflict of interest between the sexes.

Millipedes (class Diplopoda) mating on a branch.

Biologists have long been fascinated with the diversity of ways in which copulation is achieved. Research has typically focused on the means by which males and females use to locate one another and the processes of courtship, mate selection, copulation, and insemination. In addition, biologists have become interested in what happens after insemination, noting that, when females mate with multiple partners, males are selected to take whatever measures they can to ensure that their sperm supersede those of the female's other mates. Because natural selection usually works at the level of the individual, members of both sexes are adapted to behave selfishly, and Behaviors that increase the male's chances of successful reproduction, despite being detrimental to females, have arisen.

Two American rubyspot damselflies (Hetaerina americana) mating.

As a result, mating is not a simple cooperative endeavour. On the contrary, male and female interests often conflict each step of the way, from mating to allocation of parental effort. The end result of these conflicts has been an extraordinary diversity of sexual ornaments, sexual signals, genital morphology, and parental Behavior. There is, however, a diversity of solutions that range from the colourful sexual displays and elegant melodies of male songbirds to the sex-role reversal in sea horses and pipefishes (family Syngnathidae), where males carry fertilized eggs in a kangaroo-like brood pouch.

Mating interactions are usually described in terms of how many mates individuals have, how stable mating pairs or breeding groups are over time, how males and females locate one another, and how mating groups occupy space. In marine invertebrates with broadcast promiscuity, both eggs and sperm are shed into the sea to drift or swim in search of each other. Promiscuous mating, on the other hand, refers to cases in which males and females do not form long-term pair bonds and individuals of at least one sex, usually males, fertilize more than one member of the opposite sex. In promiscuous species, the sexes may meet at mating arenas or conventional encounter sites, in areas of home range overlap, or during a brief liaison in one or the other's territory. Examples include species such as the sage grouse (Centrocercus urophasianus), whose males congregate at communal display sites (leks), and a wide variety of insects species whose mating is brief and pairing is transient.

General Mating System Types

Mating System	Description
Promiscuity	Mating with multiple partners; no long-term pair bonds.
Broadcast Promiscuity	In aquatic environments, eggs and sperm drift or swim in search of one another.
Polygamy	Mating with multiple partners; stable bonds with multiple partners.
Polyandry	Females mate with multiple males.
Polygyny	Males mate with multiple females.
Resource defense polygyny	Males defend clumped resources and gain access to multiple females attracted to the resources.
Female defense polygyny	Dominant males defend a cluster of multiple females.
Scramble competition polygyny	Males compete for access to mates based on differences in their ability to move about and locate females.
Cooperative polygamy (polygynandry)	Stable breeding group made up of multiple males and females.
Monogamy	Stable, long-term male-female pair bond.

Although polygamy also involves mating with multiple partners, it often refers to cases in which individuals form relatively stable associations with two or more mates. Most such species exhibit polygyny, in which males have multiple partners. Some examples include the red-winged blackbird (Agelaius phoeniceus) and house wren (Troglodytes aedon) in North America and the great reed warbler (Acrocephalus arundinaceus) in Europe. In a few polygamous species, however, females mate with and accept care from multiple partners, a phenomenon referred to as polyandry, examples of which include spotted sandpipers (Actitis macularia), phalaropes (Phalaropus), jacanas (tropical species in the family Jacanidae), and a few human societies such as those once found in the Ladakh region of the Tibetan plateau. Monogamy, where a single male and female form a stable association, is rare in most taxa except for birds, where at least 90 percent of species are socially monogamous. Rarest of all are stable breeding groups made up of multiple males and multiple females. In such groups, all males can potentially breed with any of the females. This pattern is referred to as cooperative polygamy or polygynandry. Examples of this type of mating system include the acorn woodpecker (Melanerpes formicivorus) in western North America, the dunnock (Prunella modularis) in Europe, a few primate societies including chimpanzees (Pan troglodytes), and at least one human society, the Pahari of northern India.

The distinction between promiscuous and polygamous mating associations is a function of pair stability. In the latter, mates come together for longer than is required to fertilize eggs. Within polygamous species, however, there is considerable variation in stability. In some cases, females have one mate at a time but change mates periodically. This pattern may be referred to as serial polyandry, sequential polyandry, or serial monogamy, depending on whether the focus is on mate-switching Behavior or the number of mates at a given time. Serial monogamy can be used to describe species such as the milkweed leaf beetle (Labidomera clivicollis), in which males and females remain together for hours or days. Serial monogamy can also be used to refer to bird species such as the European house martin (Delichon urbica) and greater flamingo (Phoenicopterus ruber), in which males and females are socially monogamous within a season but acquire a new mate each year.

In contrast, simultaneously polygamous species (such as red-winged blackbirds) and simultaneously polyandrous species (such as the jacanas) also occur. Red-winged blackbird males often have two or more females breeding on their territories, whereas jacana females are bigger than males and defend large territories encompassing the smaller territories of their male mates. The distribution of these mating systems varies considerably among groups. For example, although social monogamy is common and polygamy rare in birds, the converse is true in mammals; a large fraction of mammals are polygamous. Only a handful of mammal species, including most human societies, are socially monogamous.

In addition to classification schemes based on number of mates and stability, mating associations are sometimes categorized on the basis of how individuals occupy space. Many species of songbirds defend "all-purpose" territories that provide individuals or small groups with both nesting habitat and a significant degree of exclusivity when it comes to exploiting the resources in a particular area. Other birds, particularly many seabirds, nest in colonies and defend only a small area around their nest.

The distribution of resources can influence the use of space and consequently the nature of the mating system. When females are clumped, either because of clumping of food and nest sites or because of the benefits of forming social alliances with other females, dominant males are able to defend females directly and gain multiple mating opportunities (female-defense polygyny). Alternatively, if males defend clumped resources, they can gain access to multiple fertile females attracted to the resources (resource-defense polygyny). Scramble competition polygyny is thought to occur when neither female-attracting resources nor females themselves are economically defendable. Scramble competition polygyny involves males competing for access to mates based on differences in their ability to move about and locate females. Finally, in lekking species, males aggregate at display sites that may not be tied to either resources or females. These terms focus on ways in which the ecology of space use by females influences a male's ability to monopolize mating opportunities.

Of the various kinds of mating systems, polygyny is relatively common and polyandry rare. This prevalence of polygyny is thought to result from the greater resource investment females have in their large, immobile eggs compared with males' investment in small, motile sperm.

Originally, all gametes were probably similar in size and mobility, with the defining feature being that they fused to produce a new individual. Eggs and sperm are thought to have diverged in size due to the contrasting advantages of being either small and mobile or sedentary and large. It is easiest to understand this concept by thinking of a single-celled organism that divides into two equal sex cells. Each cell contains half of the organism's genetic material. Because organisms are inherently variable, the sex cells will tend to vary somewhat in size. Assume that smaller cells move faster, thereby increasing their chances of locating another cell with which to join. In contrast, larger cells move more slowly but have more resources to devote to survival and reproduction. The increased ability of small, motile sex cells to find cells with which to fuse and the greater survival conferred upon large, slow gametes would put gametes of intermediate size and mobility at a selective disadvantage. In this context, motile cells that preferred to join with larger, more sedentary sex cells would be favoured. Consequently, gametes of intermediate size and mobility would be selected out of the population through the greater success of the two extremes. The process of selecting against intermediate individuals in favour of those individuals with extreme forms of a critical trait is known as disruptive selection.

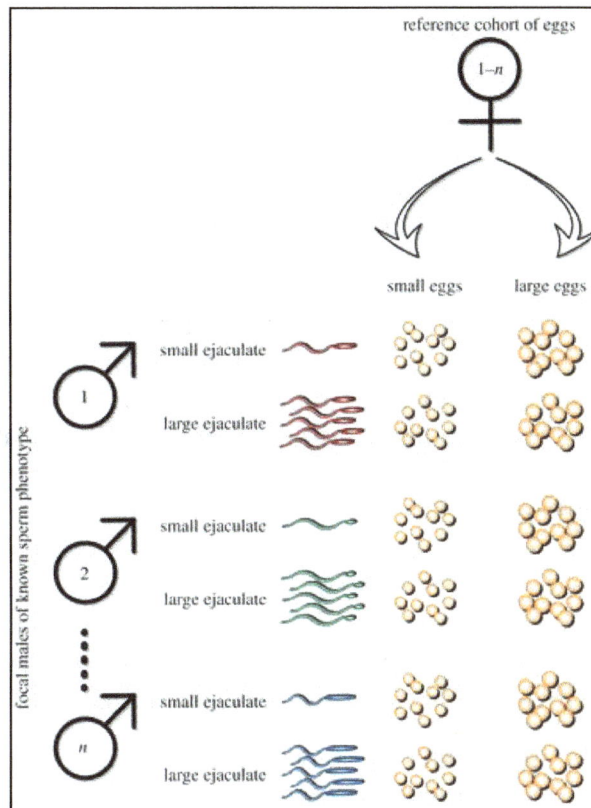

In multicellular organisms, males produce sperm, and females, which typically have a greater investment in large eggs, are usually the caretakers of eggs and young. Because males typically produce a great many relatively inexpensive sperm, they can increase the number of offspring they sire by fertilizing additional females. Thus, their reproduction is less constrained by the availability of time and resources than is female reproduction. To the extent that a male's offspring can survive without further contribution on his part, the male is free to move on and search for additional mates. Females, on the other hand, are potentially limited by time and the availability of nutrients needed to produce eggs. Unless they receive additional resources to turn into eggs, the acceptance of additional matings will not help them produce more offspring.

One consequence of this difference is that females are frequently more selective than males. There are at least three hypotheses that attempt to explain the near ubiquity of female choice.

First, females may benefit by preferring to mate with males that contribute to the physical care of offspring and thus augment the level of care their young receive or relieve females of some of their parental duties. More specifically, females should prefer males that provide resources that increase their survival and breeding success. These constitute potential "direct benefits" of mate choice.

Second, a female may choose a mate based on some apparently arbitrary male character (such as "attractiveness"). This character will allow her to produce more sons possessing that character, and these sons will ultimately attract more females and produce more grandchildren. Through a process referred to as the "sexy son hypothesis," this can result in runaway selection, a preference for exaggerated traits that are advantageous solely because of their attractiveness to females.

Runaway selection was first proposed by English statistician R.A. Fisher in the 1930s. Evidence supporting this process has been found in several species. One of the most dramatic may be the African long-tailed widowbird (Euplectes progne); the male of this species possesses an extraordinarily long tail. This feature can be explained by the females' preference for males with the longest tails, as demonstrated experimentally by artificially elongating the tails of male widowbirds. Similarly, male European sedge warblers (Acrocephalus schoenobaenus) with the longest and most elaborate birdsongs are the first to acquire mates in the spring.

In both of these cases, the traits females prefer may be arbitrary indicators of attractiveness. Alternatively, they may be most elaborately developed in males that are otherwise of high genetic quality, in which case they fall into a third possibility, where female choice is due to what is called the "good genes hypothesis." This hypothesis suggests that the traits females choose are honest indicators of the male's ability to pass on copies of genes that will increase the survival or reproductive success of the female's offspring. Although no completely unambiguous examples are known, evidence in support of the good genes hypothesis is accumulating, primarily through the discovery of male traits that are simultaneously preferred by females and correlated with increased offspring survival. For example, female North American house finches (Carpodacus mexicanus) prefer to mate with bright, colourful males, which also have high overwinter survivorship. This suggests that preference for mating with such males increases offspring survival.

The initial size asymmetry in the gametes produced by the sexes sets the stage for sexual conflict over when and with whom females mate and the amount of resources males contribute to the female and her offspring. Females may try to control the situation by choosing mates that will provide them with resources or help with parental care. They might assess males on the basis of the quality of their territory, how much food they provide during courtship, or how long a male is able to produce a particularly intricate display.

True genetic monogamy is rare. Although females do not gain in numbers of fertilizations the way males do when they mate with multiple partners, females often mate with multiple males. Why they do so is not clear. If females mate opportunistically, then happen to come across a more-preferred male, they may "trade up" in quality to increase the breeding success of their sons or the growth and performance of their offspring. Offspring performance may increase because the new male offers "good genes" or because his genes better complement those of the female. Otherwise, females may mate with multiple partners as insurance against the possibility that sperm from their first mate are inviable or in exchange for resources provided by additional males.

Multiple mating by females is not always obvious. In birds, over 90 percent of species are socially monogamous, breeding as simple pairs made up of one male and one female. Paternity tests with DNA fingerprinting, however, have revealed that females of many socially monogamous birds accept copulations from males in addition to their social mate. Such extra-pair copulations may provide females or their young with benefits. For example, female blue tits (Cyanistes caeruleus) that accept copulations with males in addition to their mates have faster-growing offspring, suggesting genetic benefits of extra-pair mating. In red-winged blackbirds, the females not only benefit through increased offspring performance, but they are allowed access to food on the extra-pair male's territory. In these cases, as both the females and their social mates feed nestlings, the male-female conflict appears to have been resolved in favour of females.

In insects and spiders, females commonly mate with multiple males. In some species, females benefit by receiving nutrients that are shunted into egg production. For example, males of certain crickets (family Gryllidae), katydids (family Tettigoniidae), butterflies, and moths (order Lepidoptera) contribute up to 25 percent of their body weight at mating, packaging their sperm in a nutritious envelope that the female consumes or absorbs. Male scorpionflies (Panorpa) hand off gifts of insect prey in exchange for copulation, saving the female the energy and risk of predation incurred by foraging for herself. Some crickets even allow females to consume their nutritious fleshy wing pads during mating and, in the most extreme cases, represented by red-back or black widow spiders (Lactrodectus), males may be partially or entirely consumed by their mates during mating.

The special form of mating competition that occurs when females accept multiple mating partners over a relatively short period of time is known as "sperm competition." The potential for overlap between the sperm of different males within the female has resulted in a diversity of behavioral adaptations and bizarre male strategies for maximizing paternity. Sperm competition, for example, is thought to be the primary reason why males offer nuptial gifts to females or allow females to cannibalize them. Such nuptial gifts are best thought of as mating effort (that is, effort directed at increasing the number of offspring a male sires) rather than parental effort, because these resources are usually not mobilized in time to benefit the offspring that are sired by the male making the donation. In addition, the male's paternity and the number of sperm he transfers often correlate with the size of the donation, suggesting that the donation functions to increase the number of offspring he sires.

Sperm competition favours the evolution of paternity guards or mechanisms for reducing the impact of sperm competition. In many animals, sperm competition results in mate-guarding Behavior, whereby males remain near the female following mating in an attempt to keep additional mates away from her prior to the fertilization of her eggs. For example, in the cobalt milkweed beetle (Chrysochus cobaltinus) the male rides on the back of the female for several hours. By engaging in this Behavior, the male sacrifices time he could use to locate a new mate in favour of preventing the female from copulating with other males before she can lay her eggs. Male damselflies and dragonflies (order Odonata) use their genitalia to physically remove or compact the sperm of the female's prior mates before inseminating her with their own sperm. In the polygynandrous dunnock or hedge sparrow (Prunella modularis), a common English backyard bird, males peck at the female's cloaca. This activity causes her to release a droplet of semen containing the sperm of prior mates before a new male begins to mate with her. In acorn woodpeckers, another polygynandrous species, the threat to a male's paternity comes from other males within the same breeding group.

As a result, males spend virtually all their time within a few metres of fertile females, guarding them from other breeder males in the group. Birdsong and territorial defense Behaviors have also been shown to function as paternity protection, although these Behaviors have other primary functions.

Courtship Behavior refers to interactions specifically directed at enticing members of the opposite sex to mate. This Behavior can involve display or direct physical contact. Historically, courtship was viewed as a mechanism of species recognition. More recently, biologists have focused on how courtship might also function in mate choice. Except in polyandrous species where sex roles are reversed, males are typically the ones that court. If females elect to mate with males with elaborate courtship signals (such as the greatly elongated tail of the male long-tailed widowbird), then this preference will be reinforced over time by the greater ability of the male offspring that possess the signal to attract mates. This preference will also be reinforced if both the courtship signal and the preference for it are inherited. After generations of successful reinforcement, the preference for the courtship signal will become common in the local population. Other populations that are physically separated from this population may not adopt this courtship signal. If this occurs, courtship Behavior may become so different that members of the local population will no longer interbreed with members of other populations. Eventually, this difference in courtship Behavior between one population and another may lead to the formation of two separate species.

The potential for rapid evolution of sexual displays due to female choice may be enhanced if females have a preexisting sensory bias to prefer a particular male trait. Examples of such biases include a preference for a lower or deeper call (in some frogs) or a long, pointed swordlike tail (in swordfishes). Once this bias is in place, any mutation that permits males to possess such a feature will be favoured and spread rapidly through the population.

Courtship can be used to mitigate danger in predatory species if there is a risk that the male will be mistaken for prey and eaten by the female. Although courtship signals are typically used before copulation to entice females to mate, they are sometimes used during copulation (copulatory courtship) to stimulate the female to accept additional sperm or after copulation (postcopulatory courtship) to improve the chance that a male's sperm will outcompete the sperm of rivals. Copulatory courtship is quite common in some species of leaf beetles (family Chrysomelidae) and appears to be related to success in spermatophore (a package or capsule containing sperm) transfer and sperm competition.

Courtship signals can be costly to produce and dangerous to bear. For example, the nocturnal trills of crickets attract parasitic flies. On the other hand, the elaborate and conspicuous displays of courtship of bowerbirds (family Ptilonorhynchidae) may be less costly than previously assumed if they are largely a function of experience. When courtship signals are costly, it is presumably difficult for males of low quality to trick females by producing signals that are as attractive as those produced by males of higher quality. Consequently, courtship Behavior is often considered an honest or reliable indicator of male quality.

Social Interactions Involving the Costs and Benefits of Parental Care

The costs and benefits of parental care will determine whether parents care for their offspring and the degree to which they are involved. Parental care is expensive in terms of both current and future costs of reproduction, which explains why the majority of animals do not care for their young. Current costs are illustrated by the example of a female guarding a clutch of eggs at the expense of laying another clutch or a male that cares for nestlings rather than attracting additional mates. An example of a future cost is the reduction in postbreeding survival suffered by willow tit (Poecile montanus) parents that fledge a large brood of offspring.

The main benefit of parental care is offspring survival, although care can also influence an offspring's condition and future reproductive success. The simplest form of parental care is guarding or protection of eggs in egg-laying, or oviparous, species. Investment in egg protection ranges from construction of an egg case to guarding exposed eggs, carrying eggs on the body surface, in a brood pouch, or in the mouth, and nest building or active nest defense. In some insects there is a continuum that ranges from laying eggs to retaining eggs inside the female's body until they hatch and are borne as larvae or live young (ovoviviparity). Parental Behavior can be extended beyond hatching or birth. Examples include treehopper females that stay with nymphs until they mature, emperor penguin (Aptenodytes forsteri) parents that feed young for several months after the eggs hatch, and human parents who frequently provide substantial parental care to their children through puberty and beyond.

In animals that provide parental care, females are generally the ones that primarily bear the costs. They spend time laying eggs, creating egg cases, guarding eggs or larvae, building nests, incubating and brooding young, carrying young (gestation), nursing (lactation), and subsequently feeding and defending offspring. Parental care by both sexes (biparental care) is much less common, however, and exclusive care by the male is rare. For example, in terrestrial arthropods, female-only care occurs in 72 orders, biparental care occurs in 13, and male-only care occurs in just 4. In addition, females in 19 orders bear live young, caring for eggs or for eggs and larvae inside their bodies.

Because parental care is costly, it is expected that a conflict of interest will arise between the sexes over whether to care for offspring and how much care to provide. Frequently one sex or the other is able to "win" this conflict by being first to abandon the offspring, leaving the remaining parent, often the female, with the choice of providing all the necessary care by herself or suffering total reproductive failure.

There are several possible reasons why males are able to abandon more frequently than females. First, because fathers lose opportunities to fertilize additional eggs by caring for young, the costs of parental care may be relatively greater for males than for females. Second, if females engage in extra-pair or multi-male mating, they will experience greater benefits of care because their share of parentage is greater than that of their social mate. For example, in an insect where females mate with multiple males and store sperm for long periods, all eggs will belong to the female, but it is unlikely that all will be sired by a single male. The lower a male's expected share of paternity, the less likely he should be to provide care for the offspring. Surprisingly, even though over 90 percent of socially monogamous birds have extra-pair fertilizations, this does not appear to result in male desertion in many cases, and sensitivity of male care to loss of paternity is uncommon.

Third, because of physiological constraints, females are sometimes more crucial for offspring survival than males. This is particularly true in placental mammals where the father can desert immediately after fertilization, often with little or no effect on offspring survival. In contrast, the mother cannot desert because she carries the offspring internally through gestation and subsequently provides essential care through lactation after birth.

Timing of gamete release could also be a factor in desertion. A testable hypothesis involves predictions of an association between order of gamete release and which sex deserts in externally fertilizing species. A researcher could then ask: Is the sex that releases gametes first more likely to desert? There is superficial support for this hypothesis to the extent that male parental care is most prevalent in fishes with external fertilization. In such fishes, males often release sperm after females release eggs. A second prediction of this hypothesis, however, is that the frequency of single-parent care by males and females should be equal in species of fishes where males and females release gametes simultaneously. This prediction is not borne out. Instead, males are significantly more likely to provide care in such species than females. Thus, the opportunity to desert does not provide a general explanation for why it is usually the females that provide care. Instead, it is possible that females give care more often because they are more likely to be close to the eggs or offspring at the time when care is required. This hypothesis predicts that males should be more likely to provide care in species whose females lay eggs immediately after copulation than in species that require a period of time between copulation and the egg-laying period. Since such delays tend to occur in fishes with internal fertilization, simple proximity to the young and the suite of factors contributing to a separation of time between fertilization and egg laying probably play important roles in determining which sex provides parental care.

Lions (Panthera leo) provide a good example of females doing the majority of parental care. Lionesses not only carry the fetus and lactate, but they perform most of the hunting for the social group, including for the larger, more dominant males. Cases in which males contribute the majority or all of the care are relatively rare; however, since these instances are so unusual, they have attracted wide attention. Well-known examples of male care include giant water bugs (family Belostomatidae), in which the female lays eggs on the male's back, and sea horses and pipefishes (family Syngnathidae), in which males carry the eggs and brood the young. Other examples include mouth-brooding frogs, fish, and various shorebirds (such as jacanas) in which females lay eggs in the nests of several incubating males. Exactly what has emancipated the females of the relatively few species with male care remains a mystery. Modern research is directed at uncovering the reasons why, in these cases, the ratio of benefits to costs for males is apparently greater than that of females.

Biparental care is almost nonexistent in insects, fish, reptiles, and amphibians. It is rare in mammals and relatively common in birds. In some species of birds with biparental care, the absence of the male results in increased or even complete nestling mortality. In other species, however, male absence has little effect. In addition, male parenting in birds may be favoured by the female's tendency to divorce males that fail to provide care or by the female's preference for males that contribute to parenting.

Some forms of parental care (such as the defense of a nest) can be shared among offspring, whereas others (such as providing food) cannot be partitioned without reducing the average offspring benefit. When parental care cannot be shared, it results in competition among siblings. If resources are scarce, offspring may compete through cannibalism, siblicide, and by directly interfering with each other's access to food, shelter, or other resources. In great egrets (Casmerodius albus), for example, the first-hatched chick typically kills its younger sibling. Younger siblings avoid this fate only in years when food is particularly abundant.

Young birds also compete for food by begging, displaying colourful gapes, or by special plumage signals to induce their parents to deliver food. Within a nest, it is often the loudest, most vigorous beggar or the chick closest to the nest cavity entrance that is fed. Use of these signals will be favoured if they help parents avoid investing in young that are weak, sickly, and less likely to survive.

Social Interactions Involving the use of Space

Although it has been established that many animals group together because it is beneficial for individuals to interact, aggregation may sometimes occur because each individual requires access to a limited resource with a patchy distribution. In such cases, clumped individuals may only appear to form a social group. In fact, each individual is exploiting the resource without interacting socially. In practice, however, the absence of interaction between individuals is difficult to demonstrate. The difficulty of distinguishing aggregations on the basis of interaction is also exemplified by some insect aggregations in which individuals communicate by using chemical or vibrational signals. Often, these signals can be detected only by using specialized equipment. Nevertheless, whether aggregations form through the attraction of individuals to one another or to a site, members experience costs that must be balanced by group benefits if aggregations are to persist.

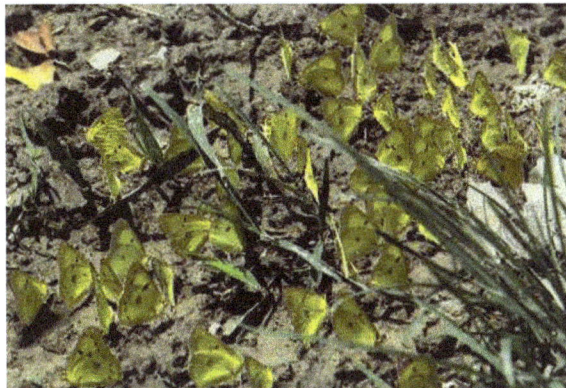

Aggregation of pink-edged sulfur butterflies (Colias interior).

The stability of aggregations is variable. Group stability ranges from temporary aggregations of bees at watering sites to gull colonies that persist on islands year after year. Among the many names used to refer to animal aggregations are covey (quail), gaggle (geese), herd (ungulates), pod (whales), school (fish), and tribe (humans) and more generalized terms such as colony, den, family, group, or pack. An even greater diversity of names is used to describe human social groups. Names such as class, congregation, platoon, squad, regiment, corps, county, town, state, and nation attest to the importance of social Behavior in virtually all aspects of human life.

The question of how aggregations form is quite different from the question of how they function. For example, use of conventional hilltop mating sites by desert butterflies is thought to involve a mutual attraction to a site, but the function of site affinity is to locate or attract a mate. Even if the proximate cause of aggregation is attraction to the site rather than to each other, this attraction to the site is thought to have arisen from benefits provided by the ultimate cause—that is, the mating opportunities the site provides.

Aggregations form for numerous reasons and in a variety of contexts. Animals benefit by forming groups when they engage in activities such as mating, nesting, feeding, sleeping, huddling, hibernating, and migrating. The plains of sub-Saharan Africa provide many examples, including lions sleeping in groups under thorn acacia trees, packs of hyenas (family Hyaenidae) cooperating to bring down a zebra (Equus quagga, E. grevyi, or E. zebra), migrating herds of wildebeest (Connochaetes), and lekking male antelopes (family Bovidae).

In order for aggregations to persist, however, the costs of group living must be balanced by the benefits. Such costs include increased competition for resources and mates, increased transmission of disease and parasites, and increased conspicuousness. Costs may increase over evolutionary time as parasites and predators evolve to take advantage of the opportunities group living provides. Nevertheless, group living also gives rise to new Behaviors that can potentially counter these increased costs. Examples of such Behaviors include nepotism (preferential treatment of kin), the formation of alliances within groups, allogrooming and allopreening (that is, activities that allow another to clean one's fur or maintain one's feathers), and communication systems that increase the benefits of group foraging and defense.

Aggregation and Individual Protection

Aggregations have been explored extensively from the standpoint of their impact on survival. The primary functions of aggregation appear to be feeding and defense. A general theory explaining why individuals should prefer to aggregate was first proposed by the Briton W.D. Hamilton, one of the most important evolutionary biologists of the 20th century. Hamilton hypothesized that animals might come together to form a so-called "selfish herd," where an individual's chances of being eaten are substantially reduced, especially if that individual remains in the interior of the group. For example, it may be better to be in the centre of a school of fish if predators tend to attack and capture fish in the outer layer. Where location within the group matters, social interactions will likely sort out social status, with some individuals gaining favoured positions by dominance or by nepotism (that is, preferential treatment shown to one's relatives).

Schoolmasters (Lutjanus apodus) on a reef in the Cayman Islands.

Living in groups also protects group members through a dilution effect. The general idea is that a predator can consume prey at only a given rate and can usually eat just one prey animal at a time. Consequently, animals in groups tend to overwhelm a predator's consumption capacity. Thus, any given individual has a smaller chance of being eaten. In the simplest example, when a group-living individual encounters a predator that will eat just one prey item, his likelihood of being eaten is reduced from p, the probability when alone, to p/N, the probability when the individual is a member of a group of size N. For example, if a tadpole joins a group with just one other individual, it reduces its chance of being eaten by one-half. Furthermore, if that tadpole joins with 99 others, its chance of being eaten drops by 99 percent. The dilution effect functions even if the group is more easily detected by predators than lone individuals are, provided that the cost of increased conspicuousness does not overtake the benefit of dilution. In other words, if the group attracts too many predators, a given individual may be better off living alone.

Alarm calls and other complex signaling Behavior within aggregations can also reduce the likelihood of predation. Calls may coordinate a group's escape from danger, confuse a predator, and prompt individuals to seek protected sites or shelter. Group members presumably benefit because the overall risk of a successful predation attempt is reduced. Alarm calls may also convey information about the type of predator and lead to the appropriate evasive Behavior. Alarm calls might even provide information regarding an individual predator's identity and habits.

A herd of caribou (Rangifer tarandus).

Alarm calling is usually considered a good example of an altruistic Behavior. Why individuals give an alarm call to begin with is not necessarily obvious, since the act of calling may attract a predator and endanger the caller. In the Sierra Nevada mountains of California, Belding's ground squirrels (Spermophilus beldingi) call more frequently when they have close relatives nearby, suggesting that alarm calling has evolved through kin selection. Alarm calls are also given by birds in flocks of mixed species and aggregations where kin selection is unlikely to be important. Such actions suggest that there are advantages of sharing the tasks associated with vigilance even in the absence of nepotism.

Group membership may also permit cooperation in defense against predators. An insect example of cooperative defense against predators is an Australian sawfly (family Pergidae); its larvae aggregate on leaves and jointly regurgitate noxious substances when attacked. A well-known mammalian example is the circle formation of musk oxen (Ovibos moschatus) in the Arctic; this arrangement serves as an effective defense against wolves (Canis lupus).

Furthermore, aggregation may augment and bolster signaling systems. This is particularly true in species with an aposematic mechanism (that is, a feature that allows a species to advertise its dangerous nature to potential predators). The grouping of aposematic prey increases the chance that a predator will have prior experience of the species, recognize the prey as distasteful, and avoid it.

Groups of animals may also confuse predators by looking larger than they actually are or by moving apart in unpredictable ways. These actions often cause the predator to hesitate just long enough to permit the prey's escape. In some beetles it is common for a male to ride on the female's back for long periods. Although this Behavior may have several costs, one possible benefit is that both the male and the female may confuse the predator; a puff of breath from the predator or its sudden movement causes the pair to separate from one another. Both individuals may have time to escape before the predator understands what took place.

Cooperative Foraging

In addition to increased vigilance and group defense, individuals in groups may benefit by cooperating to gain access to food and other resources. There is evidence that some newly hatched insect larvae overcome the physical defenses of plants better in groups than alone; they are able to enter the surfaces of leaves or pine needles more easily. In other plant-feeding insects, feeding itself affects the quality of the food. Substances in the insect's saliva that overcome chemical defenses or alter the metabolism of the host plant may allow the release of more nutrients.

When predators hunt in groups, their prey may become confused. Confusion can lead to the so-called "beater effect," a condition where prey flushed out by group activity become easy to capture. Where predators cooperate (such as in the hunting practices of lions, hyenas, and wolves), they can corner and bring down prey more easily.

Group living often selects for sophisticated systems of communication and cooperation that enhance the group's overall foraging success. For example, eastern tent caterpillars (Malacosoma americanum) follow silk-and-chemical trails. When unhomogenized milk was home-delivered in English cities, it was shown that English blue tits (Cyanistes caeruleus) could observe and learn from one another how to open the tops of milk bottles and skim off the cream.

Social Interactions Involved in Monopolizing Resources or Mates

The home range of an animal is the area where it spends its time; it is the region that encompasses all the resources the animal requires to survive and reproduce. Competition for food and other resources influences how animals are distributed in space. Even when animals do not interact, clumped resources may cause individuals to aggregate. For example, clumping may occur if individuals settle in an area one by one. Each individual weighs the costs and benefits of settling and sharing resources in high-quality areas versus settling in less dense, low-quality areas. This sort of spacing is predicted by algebraic cost-benefit models and is called the ideal free distribution. For example, if one person throws pieces of bread into a pond at twice the rate of a second person nearby on the same pond, ducks will distribute themselves between the two sources of food. The distribution will occur in approximately the same ratio as the food being provided. In other words, twice as many ducks will congregate near the person throwing the double lot of food.

Spacing patterns may occur for other reasons. Clumping may arise if individuals exhibit a mutual attraction to each other. Conversely, if individuals repel each other, they may be overdispersed (that is, more spread out and regular than would be predicted by random settlement). Social interactions that commonly influence spacing include territoriality and dominance; both are major means of monopolizing access to resources.

Territoriality

Territoriality refers to the monopolization of space by an individual or group. While territories have been defined variously as any defended space, areas of site-specific dominance, or sites of exclusive monopolization of space, they can be quite fluid and short-term. For example, sanderlings

(Calidris alba) may defend feeding territories involving a short stretch of beach during high tides, while individual male white-tailed skimmers (family Libellulidae) defend small sections of ponds as mating territories for only a few hours, effectively "time-sharing" the same area with several other males within a day. Consequently, the current approach is to view territoriality as a fluid space-use system. In this system, a resource or area is defended to varying degrees and with varying success, depending on the costs and benefits of defense.

The tendency to hold territories varies among closely related species, within species, and through time. The same individual may blink in and out of territorial Behavior as the distribution of resources, the competitive environment, or the individual's internal physiological state changes. Biologists believe that territoriality is favoured where resources are economically defendable (that is, where the benefits of restricting access outweigh the costs of defense). Costs of territoriality depend upon the energy required to keep out intruders and the potential costs of direct combat. These costs are balanced by benefits that include exclusive access to food, mates, breeding sites, and shelter.

A territory's extent varies among species. Typically, territories include sites of egg deposition, burrow entrances, nest sites, food plants, feeding space, advertisement perches or display sites, roosting sites, shelters, grazing areas, food stores or communal caches, foraging space, and even patches of sunlight in the forest. Territories may contain a single critical resource, such as the bee nests defended by male orange-rumped honey guides (Indicator xanthonotus) in the Himalayas. In other cases, as in many territorial songbirds, males defend multipurpose territories for which it is difficult to identify a single key resource.

The costs and benefits of competing for space, and ultimately resources, depend on the density of competitors and on how resources are distributed. When resources are clumped, they are more easily managed and defended. In contrast, as they become increasingly spread out or as their relative quality declines, the benefits and ease of defense are reduced. Conversely, when resources are too high in quality, competition may be so intense that exclusivity is impossible or simply too costly to maintain. Consequently, territoriality is generally expected when resources are of intermediate quality.

If the quality of a resource varies by season, there may be periods when the resource no longer provides enough benefits to warrant defense. If this is true, territoriality should correspond to the period of greatest benefit. For example, Yarrow's spiny lizards (Sceloporus jarrovii) appear to maintain mating territories only when the majority of females are receptive to mating. As more preferred areas are taken, some individuals forgo territoriality. In rufous-collared sparrows (Zonotrichia capensis), for example, males without access to high-quality territories live on the fringes of the territories of older, more dominant males.

Dominance

Territoriality is one way that animals compete for and partition resources. Within groups, individuals may compete for resources and space by means of social dominance. Dominance interactions refer to the Behaviors occurring within or between social groups that result in hierarchical access to resources or mates; they do not refer to the use of space. Dominant individuals are characterized as being more aggressive and successful in winning competitive interactions than other group members. Dominance may be established through direct or indirect aggression or by mutual display, where the dominant individual usually assumes a higher stature and the subordinate often bows or mimics juvenile Behavior.

Payoff* to...	...in fights against:	
	hawk	dove
hawk	Hawk wins 50% of fights; is injured in 50% of fights. Payoff: $(V-D)/2$	Hawk always wins; dove flees. Payoff: V
dove	Dove never wins; is never injured. Payoff: 0	Dove wins 50% of fights; is never injured; wastes time. Payoff: $V/2-T$

*V = fitness value of winning resources in fight
D = fitness costs of injury
T = fitness costs of wasting time

As with many other aspects of social Behavior, an economic argument is used to explain why dominance is sometimes resolved by display rather than fighting. Because symmetrical contests involve contestants that by definition have an equal chance of winning, contests involving individuals close in dominance status should involve the most fighting. In contrast, when one individual is clearly superior, the lesser individual will gain little by challenging and may even suffer injury in the process of trying. Thus, clearly established dominance hierarchies are thought to be advantageous to both dominants and subordinates due to a reduction in the frequency of energetically expensive and dangerous fighting. Often, life is smooth within social groups not because of a lack of competition, but because dominance is established and the hierarchy is clear.

Dominance hierarchies have been shown to play a critical role in mating patterns in black-capped chickadees (Poecile atricapillus), where more dominant males tend to mate with more dominant females. Higher-status pairs then experience greater overwinter survival, presumably compete more effectively for high-quality breeding space, and produce more offspring.

Dominance often correlates with mating success in polygynous societies. In some cases, dominant males gain preferred positions in mating arenas and are more likely to be chosen by females. An understanding of why subordinates should accept their lower-status can be gained by examining the options available to lower status individuals. A subordinate has a finite number of choices: remain in its social group, join another group where its chances are better, or become solitary. Solitary individuals will lose the benefit of being in a group, and individuals that emigrate will face the difficulties of locating and joining a new group. If the new group offers greater opportunities for achieving high status, emigration will be favoured. Familiarity with group members and with foraging and shelter sites will favour remaining with the group. The future opportunities of young animals may be enhanced by the skills they learn as subordinates, and, when groups comprise relatives, nepotism may also favour staying. Often, subordinates are willing to bear the costs of reduced access to mates and resources when the alternatives available to them are even worse.

Subordinates often exhibit an array of tactics or Behaviors that help them make the best of their low status. These alternative strategies include the sneaky mating tactics of subordinate male bullfrogs (Lithobates catesbeianus) and the specialized group of small male ("jack") coho salmon (Oncorhynchus kisutch), which act as "satellites" and try to intercept females as they are attracted to the territories of large males. Other examples include the female-mimicking Behavior of subordinate male rove beetles (family Staphylinidae) and the satellite Behavior of horseshoe crab (Limulus polyphemus) males. In the former example, mimicks benefit from reduced aggression and thus increased access to matings; in the latter, subordinate male horseshoe crabs may fertilize some of a female's eggs while she is mating with a more dominant male. Such alternative reproductive tactics enable males to circumvent the constraints of low status. In some cases, these activities may allow subordinate males to achieve fitness benefits comparable to those of more dominant individuals.

Social Interactions Involving Movement

The benefits of forming dispersal swarms, flocks, and coalitions are considered similar to the advantages of living in aggregations as both exploit the potential benefits of living in groups. Moving about in groups can provide additional advantages, such as the reduction in turbulence and energy savings accrued by geese migrating in V-formations. However, dispersal and migration are energetically expensive and fraught with danger because they require facing unfamiliar surroundings.

If group size is associated with the ability to compete for and monopolize space, specialized breeding areas, or wintering sites, group dispersal may yield advantages when it comes time to settle. For example, increased group size makes coalitions of lions and coalitions of acorn woodpeckers more competitive in fights for the infrequent breeding vacancies arising in other groups. In the case of lions, however, these benefits do not extend to the female prides for which the males compete; males often kill unrelated infants upon joining a pride to increase their own chances of siring offspring with the group's females.

Social Interactions Involving Cooperative Breeding and Eusociality

Cooperative breeding occurs when more than two individuals contribute to the care of young within a single brood. This Behavior is found in birds, mammals, amphibians, fish, insects, and arachnids; however, cooperative breeding is generally rare because it requires parental care, which is itself an uncommon Behavior. In birds, which have a high taxonomic commitment to biparental care, about 3 percent of species are cooperative breeders. Cooperative breeding is generally linked to cases of restricted dispersal and cases where opportunities for prolonged contact between close relatives occur (such as in species inhabiting mild climates with year-round residency).

In vertebrates, most cases of cooperative breeding involve helpers at the nest (such as offspring from prior years that remain near their parents and help rear younger siblings). Species with helpers include common crows (Corvus brachyrhynchos), Florida scrub jays (Aphelocoma coerulescens), and a variety of tropical species—particularly in Australia. Relatively few cases involve cooperative polygamy or mate sharing, in which there are multiple cobreeders of one or both sexes.

Examples of mate-sharing Behavior occur in acorn woodpeckers (Melanerpes formicivorus), dunnocks (Prunella modularis), and common moorhens (Gallinula chloropus).

The outcome of mate sharing in birds and other taxa where reproduction is potentially shared is highly variable. In so-called egalitarian societies, two or sometimes three breeders may share maternity equally (as occurs in joint-nesting female acorn woodpeckers). In contrast, in some societies reproduction is highly biased toward the activities of a single individual (frequently referred to as "reproductive skew" or "skewed reproduction"). For example, in some ant colonies a single female (the queen) lays all the eggs.

Reproductive sharing is costly and occurs in a variety of organisms. Cooperation and competition over shared reproduction may even occur in simple multicellular organisms, such as the "social amoeba" (Dictyostelium discoideum). Clones of Dictyostelium form a multicellular fruiting body called a plasmodium. Superficially, the plasmodium resembles a slug, but it is essentially an aggregation of free-living, haploid, amoeba-like, cells that will later grow a sterile stalk. The stalk raises the spores off the ground and facilitates their dispersal. Sometimes cells that come from different clones cooperate to form the plasmodium. When slugs form from two different haploid cells, the clones do not contribute equally to the reproductive spores; often a "cheater" can be identified that contributes proportionally more to spores than to the sterile stalk.

In birds, cooperative breeding is generally believed to be a result of a shortage of high-quality territories or mates, and helpers will typically become breeders if given the opportunity to do so. These constraints favour philopatric individuals (that is, those individuals who do not disperse). Those individuals stay home, where they may augment their fitness by helping their parents raise younger offspring. In some cases, there is good evidence that young birds weigh the inclusive fitness benefit of staying home and helping against the fitness benefits of settling in available, lower-quality territories. Helpers often behave parentally by feeding nestlings and defending the nest. They may vary in how much they feed or defend, but the division of labour is neither extreme nor does it tend to be fixed or stereotyped.

In Kalahari meerkats (Suricata suricatta), breeding individuals of both sexes live in cooperative groups, with dominant members accounting for the bulk of reproduction. Group augmentation, a positive group-size effect on reproduction, arises because helpers enhance pup growth and survival by babysitting, which is only done by subordinates. Babysitting sometimes involves remaining in the burrow without food for up to 24 hours. The sacrifice of helpers is measurable as weight loss, but helpers of both sexes have been shown to benefit from living in the group with fitness gains through both direct reproduction and the raising of nondescendant kin. Female subordinates become pregnant, albeit less successfully than dominants, and compete for reproductive success within the group by committing infanticide. Male subordinates have been shown to foray to other groups, where they compete to sire extragroup young. While these strategies are not equivalent to breeding as a dominant, they provide young animals with fitness-enhancing options in a breeding environment constrained by food, predation, and availability of breeding vacancies.

Eusocial insects show more extreme forms of sociality with a reproductive division of labour in which individuals form castes that perform different colony functions. The classic example of this phenomenon is the honeybee (Apis mellifera) colony. The colony is made up of a single large queen, who lays eggs, and tens of thousands of workers, who perform the work associated with foraging and colony maintenance. Similarly, in some species of termites, queens become so large with eggs that their abdomens are stretched to several times the normal body length. Their enormous size renders them virtually immobile.

Honeybee workers are effectively sterile daughters with reduced ovaries that only occasionally lay unfertilized eggs which develop into males. Workers start out by tending eggs and larvae and by defending the colony. As they age, they switch to foraging outside the hive, a dangerous task that requires navigational ability and spatial memory. Termites and ants also have workers that tend to the queen and perform colony tasks. In addition, some termite, ant, and aphid species have specialized soldier castes that are designed for defense.

Throughout the eusocial insects, there is a tremendous bias in reproduction favouring one or a few individuals and a great deal of self-sacrifice on the part of workers. Most workers will never have the opportunity to reproduce. Multiple queens occur in some social insects like paper wasps (Polistes), in which one to three females will found a colony together and share reproduction to a greater or lesser extent.

The important advance of kin selection theory as proposed by W.D. Hamilton was that individuals have an inclusive fitness that combines kin-selected fitness benefits with direct reproductive benefits into a single measure of "offspring equivalents." Normally, sisters have half their genes in common, and individuals who help parents produce an additional sister gain as much inclusive fitness as if they had an offspring of their own. What intrigued Hamilton is that certain insects of the order Hymenoptera, particularly ants, bees, and wasps, have a bizarre genetic system called haplodiploidy.

Under this system, males are derived from unfertilized (haploid) eggs with half the number of gene copies of a normal fertilized (diploid), female-destined egg. This means that haploid fathers have only one set of genes to give their daughters and that all of their sperm are identical. Diploid mothers, however, produce a multitude of genetically different eggs by assorting half their genes into eggs at random. In a group of sisters with a common father, the genes they receive from their mother are 50 percent identical, whereas all the genes they receive from their father are 100 percent identical. The result is that ant, bee, and wasp sisters share 75 percent of their genes through common ancestry, whereas they share only 50 percent of their genes with their own daughters.

In other words, because of haplodiploidy, full sisters are worth 1.5 offspring equivalents, and female workers potentially transmit more copies of their genes by helping their mother produce more sisters than by producing their own daughters and sons. This result excited Hamilton because it provided a potential explanation for why social hymenopterans often have large, apparently altruistic colonies with large numbers of workers that forgo their own reproduction to help their mother (the queen) produce more sisters. Additional study has revealed that this bizarre genetic system may be a predisposing, rather than a causal, factor in the evolution of eusociality. There is evidence, for example, that haplodiploidy is unlikely to be an exclusive cause of social Behavior in the Hymenoptera. Queens regularly mate with multiple males, and thus sperm is provided by more than one source, thereby diluting the haplodiploidy effect on sister relatedness. In addition, multiple queens may found wasp colonies, and each foundress may help to raise nieces instead of sisters.

The most widely accepted explanation for the extreme social Behavior seen in eusocial insects and mole rats is a more generalized form of kin selection combined with a reduction in opportunities for personal reproduction. Declines in personal reproduction are thought to result from high predation rates, a shortage of available nest sites, and a short breeding season. As in the case of cooperatively breeding birds, opportunities to survive and reproduce away from the colony are limited, favouring individuals that stay home. If individuals remain in their natal groups, within-colony relatedness will be high, in general, and kin selection will be a potentially important evolutionary force that favours cooperation.

Once individuals live in eusocial colonies, the selection for traits that improve colony efficiency will be strong, whereas the selection for survival of individual workers will be weak. This type of colonial living can lead to the evolution of suicidal Behavior. For example, a worker honeybee may sting a predator and die leaving its sting lodged in the victim. Hamilton's rule provides an explanation of why this and other self-sacrificial Behaviors might evolve in social species. As colony size increases, a honeybee worker's survival becomes proportionally less important to her own inclusive fitness (that is, the sum total of her ability to pass on her genes or the genes of close relatives to the next generation) than the survival of the colony.

Social Interactions Involving Communication

Communication plays a critical role in aggregation, reproductive Behavior, territoriality, dominance interactions, parental care, and cooperative interactions within families. By definition, communication involves at least one sender producing a signal conveying information that in some way alters the response of the receiver. Signaling systems are favoured when sender and receiver both gain from the interaction.

When individuals advertise their strength or condition, costly signals are favoured, because they more honestly convey individual quality. When signals are deceptive, an evolutionary arms race ensues, favouring receivers that disregard dishonest signals and senders that are increasingly deceptive. It is generally less costly to receive a signal than to send one, but receivers may also incur costs when discriminating among and responding to signals.

Signals exhibit extraordinary diversity and may involve specialized plumage, elaborate morphological characters, vocalizations, pheromones, vibrations, or chemicals that are perceived by taste. Like most adaptations, signals are usually modifications of previously existing structures or Behaviors. For example, Behaviors such as preening and feeding have become increasingly ritualized to function as signals in certain groups of animals. In many cases, displays appear to involve redirected, ritualized aggression, during which individuals compete for dominance (and thus indirectly for access to mates or resources) via contests of strength or endurance. Contestants appear to avoid using deadly force, even though in some species—such as wolves and rattlesnakes (Crotalus)—individuals appear well equipped to kill or significantly harm each other. In others, signals may have functioned originally in species recognition but were modified later to convey information about the relative quality of individuals within a species. In general, signals of mate attraction will be shaped both by the mating advantages they confer and by the advantages of avoiding the costs of hybridization.

By tracing the evolutionary history of a group of organisms, it is sometimes possible to examine how signals have evolved. For example, pheromones used by herbivorous insects may have originated with the use of plant compounds. Later evolved species produced a synthesis and a blending of chemicals that generated increasingly complex and informative mixtures. In some frogs, a preference for certain components of the male's call occurred in the ancestor of species producing the call. This modern preference suggests that the call was favoured by a preexisting bias in ancestral females.

Signals are often special modifications of starting material that either had no function or previously functioned in an entirely different context. For example, insects often produce song by stridulating (that is, rubbing body parts together). The structures used are legs and wings, although signaling in many crickets and katydids is enhanced by special rasplike modifications of the cuticle.

The breeding plumage, display Behavior, and elaborate vocal Behavior of male birds are energetically costly to produce and maintain, suggesting that they are honest indicators of age, status, and condition. Such signals also typically increase the conspicuousness of the sender. In the cases where species use elaborate signals (such as in the long tails of male African widowbirds), the ability to use a structure for its original function (flight and balance) may be compromised. In widowbirds, flight and balance costs are countered by benefits related to the female's mating preference for long-tailed males. Another classic example of a costly signal is the chuck call of the túngara frog (Physalaemus pustulosus). Females prefer the chuck call; however, by producing the call, males increase their risk of predation by bats.

The honesty of signals produced by widowbirds and Túngara frogs is maintained because only superior individuals can bear the costs of reduced flight performance or greater conspicuousness to predators. In some cases, bright plumage in male birds appears to be an honest signal of disease resistance through its complex relation to the endocrine and immune systems. Bright plumage is associated with high testosterone levels; however, testosterone itself appears to suppress the immune system. In the superb fairy wrens (Malurus cyaneus) of Australia, males vary considerably in timing of their nuptial molt, and females prefer males that molt into bright plumage earlier in the season. As a result, it is possible that only the fittest males can afford the immunity costs of maintaining bright plumage, and females might prefer bright males because they are better able to resist disease and pass on to their offspring copies of genes for resistance.

The design of a signal depends upon its function and the type of information it conveys. Function will dictate how far the signal must travel, whether or not it should convey information about an animal's location, how persistently the signal is given, the signal's variability, and how informative or arbitrary the signal is. Design will differ along these lines depending on whether it is used in mate attraction, courtship, territorial defense, aggression, or alarm. Signal evolution is also influenced by costs. For example, mate attraction signals are often highly conspicuous, whereas alarm calls are often simple tones that are difficult to locate. Signal costs can be greatly increased when other species evolve the ability to "eavesdrop" on the signaling animal. For example, the tachinid fly (family Tachinidae) may cue in on a male cricket's song and lay a parasitic egg on the cricket while he is busy attracting a mate.

The Proximate Mechanisms of Social Behavior

The proximate causes of social Behavior include the underlying genetic, developmental, physiological (that is, neural and endocrine), and morphological mechanisms. Proximate mechanisms are required to trigger the onset of a particular Behavior—such as sexual Behavior in rats (Rattus), the development of singing Behavior and song recognition in white-crowned sparrows (Zonotrichia leucophrys), the cessation of brood care and the onset of foraging Behavior in worker honeybees, and the development of bright plumage and sexual display in the superb fairy wren. While proximate mechanisms do not explain the evolutionary basis of a Behavior, they provide insight into the ways in which organisms are adapted to perform remarkably intricate and complex functions.

Early on, researchers debated the relative importance of "nature," or genetic predisposition, and "nurture," or environment, in the development of Behavior. Through extensive observation and experimentation, biologists have come to recognize that the argument is futile. Ultimately, both are important, and the interesting questions lie in how genetic predisposition and the environment interact. The environment includes such factors as nutrition, the animal's hormones, its experience of the outside world, and various features of the social milieu. Examples of the interplay between nature and nurture in the development of social Behavior can be found in studies of the inheritance of IQ in humans, song type and song learning in birds, performance of specialized tasks in eusocial species, and the mate and kin recognition systems of animals.

An excellent example of a genetic predisposition comes from studies of the migratory Behavior of blackcap warblers (Sylvia atricapilla) in Europe. When reared in captivity, the directional orientation of warblers from southwestern Germany is southwest as they begin their migration, whereas birds from Austria orient west. When the German and Austrian birds hybridize, the orientation of their offspring is intermediate between the preferred directions of the parents. The resulting change in orientation demonstrates a genetic role in determining the direction of migration.

Other Behaviors for which a clear genetic basis has been established include the dichotomy between roving and sedentary foraging in fruit flies (families Trypetidae and Drosophilidae) and the maternal Behavior of mice (family Muridae). In genetic crosses of great tits (Parus major), genetic effects accounted for variation in individual boldness and the tendency to explore new environments. In addition, newly available molecular genetic techniques have begun to generate considerable information on how genes influence behavioral development. In honeybees the switch from working in the hive to foraging is associated with a 39 percent change in gene product expression in the brain, indicating that developmental change is associated with changes in gene regulation. Behavioral genetics is a growing field with significant potential for uncovering new information on the relative inputs of nature and nurture to behavioral development.

Some of the most widely recognized evidence for the inheritance of Behavior comes from comparison of identical and fraternal twins reared apart. Identical twins come from a single egg and are genetically identical, whereas fraternal twins develop from separate eggs and share only half their genes by common inheritance. When raised in separate homes, identical twins are far more similar to each other than are fraternal twins, indicating that a variety of Behaviors and preferences have a genetic basis. One such twin study suggested that 70 percent of the variation in IQ in the study population had a genetic basis, although environmental variables (such as early nutrition and opportunities for early learning) still played important roles. Specific alleles of the dopamine receptor gene (DRD2) are associated with susceptibility to post-traumatic stress disorder and alcoholism, and this gene also appears to influence children's resilience to stress and family trauma.

Hormones, developmental mechanisms, neural mechanisms, learning, the social environment, and the physical environment all exert proximate influences on Behavior. The development of birdsong provides examples of several of these. The songbird brain has two main neural pathways. The first is a motor pathway involved in song production, and the second is a pathway in the anterior forebrain that is involved in song learning and recognition. In some species, learning is restricted to the first year of life. In others, learning is open-ended and continues long after the first year. By using a technique that destroys brain cells, biologists have been able to narrow down and identify parts of the songbird brain that are differentially involved in learning and recognition. Compared with crickets, which produce a highly stereotyped species-specific call without ever hearing another cricket, young birds must hear and practice the songs of conspecifics. The experience of hearing and practicing provides the necessary link between the auditory and song systems required to sing properly as adults. Most authorities contend that early song memories are stored in the brain as a "song template." The bird then refers to the song template as it practices the song.

The song of the zebra finch (Taeniopygia guttata) illustrates the hormonal influences on song development and singing Behavior. After the birds hatch, male and female brains develop differently. Injecting females with estrogen early in development causes them to develop malelike brains, but they will not sing male song unless they receive an implant of the male hormone, testosterone. In this example, both the early injection of estrogen and an implant of testosterone later on are necessary to produce females that sing male song. The estrogen allows females to develop the neural circuitry to undergo the learning required to produce song and the testosterone is required to stimulate females to sing.

Hormones can be used to stimulate females to exhibit malelike Behavior in other organisms. Female rats injected with testosterone as newborns will exhibit male copulatory Behavior as adults, and males castrated at birth will develop femalelike brains and Behaviors. Hormones can even be transferred among fetuses. For example, fetal mice that develop in the uterus between two males will be more aggressive later on than mice that develop between two females.

In honeybees, juvenile hormone (an insect developmental hormone) primarily influences larval development. Juvenile hormone also affects adult Behavior by stimulating development of a brain region known as the mushroom bodies. In addition, this hormone causes workers to cease brood care and begin foraging. Mushroom bodies are thought to be involved in spatial memory, an ability that enables an animal to use landmarks during trips to favoured foraging sites.

The critical importance of social influences on behavioral development can be seen throughout the period of song learning in song sparrows (Melospiza melodia). There is a sensitive period in the first summer of life when young birds learn much of their song, but field studies show that learning also continues through the first year. In song sparrows this involves developing and storing fairly exact copies of older neighbours' songs in a region of the brain called the forebrain song nuclei. A young song sparrow occupying a territory learns the songs of his near neighbours and then strings together elements from several to produce his own song. Males are more likely to store and learn song types that are shared among two or more of their neighbours. The end result is that each song sparrow holds roughly half its eight to nine song types in common with its neighbours. The adaptive function of such song sharing Behavior may be that it facilitates the rapid detection of intruders.

Mechanisms of recognition are essential if individuals are to discriminate members of their social group, choose a mate of the appropriate sex, locate their parents, care for the right offspring, and offer preferential treatment to kin. Early work in this area involved precocial birds, which often forage shortly after hatching, creating a need for mechanisms that allow them to recognize their parents. Austrian zoologist Konrad Lorenz, one of the fathers of ethology, demonstrated that gray-lag geese (Anser anser) imprint on their mothers shortly after hatching. When goslings imprinted on Lorenz, they followed him around just as they would their mother. After these geese became adults, they even courted human beings.

Sex-recognition mechanisms show imprinting effects as well. In a bill-painting experiment, young albino zebra finches were exposed to parents with bills painted different colors. After the finches became adults, males could be tricked into courting the wrong sex by reversing male and female bill colours of adults.

Parents engaging in parental care also require mechanisms that permit them to recognize their offspring. Offspring recognition probably involves odour in most insects and mammals. When they recognize their offspring, birds tend to use markings or vocalizations rather than scent. In many species of birds, parents do not recognize nestlings of other pairs that are artificially fostered into their nests; this lack of recognition is probably due to the fact that nestlings do not move around from one nest to another in the wild. Consequently, there has been no selection for such recognition. Conversely, nestlings of colonial bank swallows (Riparia riparia) often move between adjacent holes, and parents are able to recognize their own chicks on the basis of their vocalizations. In addition, offspring recognition can be extraordinarily precise as in Mexican free-tailed bats (Tadarida brasiliensis). In this species, mothers are 70 percent accurate in picking out their own pups from among thousands of pups huddled in a small area of cave ceiling.

Kin recognition systems also play a role in contexts where it pays to favour close over distant kin. The three mechanisms of kin recognition are the use of environmental cues, prior experience, and phenotype matching (that is, looking or smelling right). Examples can be found in the joint-nesting Behavior of paper wasps and the kin-directed alarm calls of ground squirrels.

Paper wasp foundresses pick up odours and odour preferences from their natal nests that are later used to discriminate and preferentially associate with nest mates when it is time to found a colony. In contrast, recognition in Belding's ground squirrels involves a combination of prior association and phenotype matching. Unrelated Belding's ground squirrels reared together treat each other as kin, whereas siblings reared apart do not. There is a component of kin recognition in ground squirrels, however, that must be based on genes or contact between individuals in the womb. Even though they have been reared apart, siblings are more likely than nonsiblings to associate with each other. The most compelling evidence for phenotype matching comes from house mice (Mus musculus): they use odour cues associated with genetic variation in the major histocompatibility complex (MHC) to recognize and avoid mating with relatives. (The MHC is a group of genes that code for proteins found on the surfaces of cells that help the immune system recognize foreign substances.)

Evolutionary Psychology and Human Behavior

Understanding the ultimate and proximate causes of social Behavior in various animals provides a compelling case that evolutionary history, natural selection, development, endocrine and neural mechanisms, and the social environment all might well affect the expression of social Behavior in human beings. The process of explaining human Behavior, however, is a daunting exercise. If songbird social and sexual Behavior is the complex outcome of a large number of developmental and physiological processes, then it is unlikely that simplistic approaches to understanding human Behavior will be accurate.

For example, American psychologist John Money considered the social environment to be of over-riding importance in gender identity. In treating children whose sex was ambiguous at birth, such as those with underdeveloped genitalia, he recommended that they be raised according to the parents' initial perceptions of the child's gender rather than the actual genetic sex of the child. This resulted in parents raising genetic boys as girls and vice versa, often requiring surgery and hormonal treatments later on in life. The complex nature of sexual Behavior observed in various animal studies and studies of human adults and adolescents suggests that Money's social approach to sexual identity has a weak scientific foundation. These studies also suggest that decisions regarding how to treat cases of ambiguous gender should instead be based on a comprehensive knowledge of the neural, hormonal, physiological, genetic, and social bases of sexuality.

Similarly, the relatively new discipline of evolutionary psychology can easily go too far in extending evolutionary principles to human Behavior. Extant behavioral traits in humans were not shaped by the current environment. Rather, the environmental context in which humans evolved was probably quite different from that of the modern world. People in ancestral societies lived in smaller groups, had more-cohesive cultures, and had more stable and rich contexts for identity and meaning. As a result, it is important to be cautious when using present circumstances to discern the selective bases of human Behavior. Despite this difficulty, there have been many careful and informative studies of human social Behavior from an evolutionary perspective. Infanticide, intelligence, marriage patterns, promiscuity, perception of beauty, bride price, altruism, and the allocation of parental care have all been explored by testing predictions derived from the idea that conscious and unconscious Behaviors have evolved to maximize inclusive fitness. The findings have been impressive. As with other species, however, it is important to critically evaluate and avoid overextending the evidence.

One of the key criticisms of human sociobiology is borne of fear that the findings will be used to effect unfair or immoral policies. Examples include use of social Darwinism to justify discriminatory practices, economic policies that benefit relatively few at the expense of many, genocide, eugenics, and legal systems that fail to protect the vulnerable segments within populations. These potential problems suggest the need for deep ethical consideration of the implications of evolutionary psychology. Such an approach would investigate how results might be used ethically, to benefit society, or unethically, to cause harm.

For example, consider the finding that stepfathers are more likely than biological fathers to abuse and kill nonbiological children in the household. This finding could be used to justify an increase in the social services available to blended families, particularly those in which the mother has biological children from a previous marriage. In developing countries, an understanding of the evolutionary context of bride-price (monetary or other resources given to the family of a potential bride) could be used to predict and ameliorate the detrimental impacts of rapid social change that accompanies an influx of Western culture and technology.

The real danger lies not in the scientific findings of evolutionary psychology but in the failure to recognize that scientific findings should never dictate ethics and morality. Policies that affect the rights, opportunities, and dignity of human beings occur within the moral rather than the scientific realm of human endeavour.

TERRITORIAL BEHAVIOR

Territorial Behavior, in zoology, is the method by which an animal, or group of animals, protects its territory from incursions by others of its species. Territorial boundaries may be marked by sounds such as bird song, or scents such as pheromones secreted by the skin glands of many mammals. If such advertisement does not discourage intruders, chases and fighting follow.

Territorial Behavior is adaptive in many ways; it may permit an animal to mate without interruption or to raise its young in an area where there will be little competition for food. It can also prevent overcrowding by maintaining an optimum distance among members of a population. Territories may be seasonal; in many songbirds the mated pair defends the nest and feeding area until after the young are fledged. In communally nesting birds such as gulls, the territory may simply consist of the nest itself.

Wolf packs maintain territories in which they hunt and live. These areas are aggressively defended from all non-pack members. The male cougar has a large territory that may overlap the territories of several females but is defended against other males. Responding to scent marks, the inhabitants of the overlapping ranges also avoid each other, except for breeding.

ANIMAL SEXUAL BEHAVIOR

Anatomical structures on the head and throat of a domestic turkey. 1. Caruncles, 2. Snood, 3. Wattle (dewlap), 4. Major caruncle, 5. Beard. During sexual behavior, these structures enlarge or become brightly colored.

Animal sexual behavior takes many different forms, including within the same species. Common mating or reproductively motivated systems include monogamy, polygyny, polyandry, polygamy and promiscuity. Other sexual Behavior may be reproductively motivated (e.g. sex apparently due to duress or coercion and situational sexual Behavior) or non-reproductively motivated (e.g. interspecific sexuality, sexual arousal from objects or places, sex with dead animals, homosexual sexual Behavior, and bisexual sexual Behavior).

When animal sexual Behavior is reproductively motivated, it is often termed mating or copulation; for most non-human mammals, mating and copulation occur at oestrus (the most fertile period in the mammalian female's reproductive cycle), which increases the chances of successful impregnation. Some animal sexual Behavior involves competition, sometimes fighting, between multiple males. Females often select males for mating only if they appear strong and able to protect themselves. The male that wins a fight may also have the chance to mate with a larger number of females and will therefore pass on his genes to their offspring.

Historically, it was believed that only humans and a small number of other species performed sexual acts other than for reproduction, and that animals' sexuality was instinctive and a simple "stimulus-response" Behavior. However, in addition to homosexual Behaviors, a range of species masturbate and may use objects as tools to help them do so. Sexual Behavior may be tied more strongly to establishment and maintenance of complex social bonds across a population which support its success in non-reproductive ways. Both reproductive and non-reproductive Behaviors can be related to expressions of dominance over another animal or survival within a stressful situation (such as sex due to duress or coercion).

Stags fighting while competing for females – a common sexual Behavior.

Greater sage-grouse at a lek, with multiple males displaying for the less conspicuous females.

Mating Systems

In sociobiology and Behavioral ecology, the term "mating system" is used to describe the ways in which animal societies are structured in relation to sexual Behavior. The mating system specifies which males mate with which females, and under what circumstances. There are four basic systems.

The Four Basic Mating Systems

	Single female	Multiple females
Single male	Monogamy	Polygyny
Multiple males	Polyandry	Polygynandry

Monogamy

Monogamy occurs when one male mates with one female exclusively. A monogamous mating system is one in which individuals form long-lasting pairs and cooperate in raising offspring. These pairs may last for a lifetime, such as in pigeons, or it may occasionally change from one mating season to another, such as in emperor penguins. In contrast with tournament species, these pair-bonding species have lower levels of male aggression, competition and little sexual dimorphism. Zoologists and biologists now have evidence that monogamous pairs of animals are not always sexually exclusive. Many animals that form pairs to mate and raise offspring regularly engage in sexual activities with extra-pair partners. This includes previous examples, such as swans. Sometimes, these extra-pair sexual activities lead to offspring. Genetic tests frequently show that some of the offspring raised by a monogamous pair come from the female mating with an extra-pair male partner. These discoveries have led biologists to adopt new ways of talking about monogamy; According to Ulrich Reichard (2003):

Social monogamy refers to a male and female's social living arrangement (e.g., shared use of a territory, Behavior indicative of a social pair, and/or proximity between a male and female) without inferring any sexual interactions or reproductive patterns. In humans, social monogamy takes the form of monogamous marriage. Sexual monogamy is defined as an exclusive sexual relationship between a female and a male based on observations of sexual interactions. Finally, the term genetic monogamy is used when DNA analyses can confirm that a female-male pair reproduce exclusively with each other. A combination of terms indicates examples where levels of relationships coincide, e.g., sociosexual and sociogenetic monogamy describe corresponding social and sexual, and social and genetic monogamous relationships, respectively.

Whatever makes a pair of animals socially monogamous does not necessarily make them sexually or genetically monogamous. Social monogamy, sexual monogamy, and genetic monogamy can occur in different combinations.

Social monogamy is relatively rare in the animal kingdom. The actual incidence of social monogamy varies greatly across different branches of the evolutionary tree. Over 90% of avian species are socially monogamous. This stands in contrast to mammals. Only 3% of mammalian species are socially monogamous, although up to 15% of primate species are. Social monogamy has also been observed in reptiles, fish, and insects.

Sexual monogamy is also rare among animals. Many socially monogamous species engage in extra-pair copulations, making them sexually non-monogamous. For example, while over 90% of birds are socially monogamous, "on average, 30% or more of the baby birds in any nest are sired by someone other than the resident male." Patricia Adair Gowaty has estimated that, out of 180 different species of socially monogamous songbirds, only 10% are sexually monogamous.

The incidence of genetic monogamy, determined by DNA fingerprinting, varies widely across species. For a few rare species, the incidence of genetic monogamy is 100%, with all offspring genetically related to the socially monogamous pair. But genetic monogamy is strikingly low in other species. Barash and Lipton note:

> "The highest known frequency of extra-pair copulations are found among the fairy-wrens, lovely tropical creatures technically known as Malurus splendens and Malurus cyaneus.

More than 65% of all fairy-wren chicks are fathered by males outside the supposed breeding group."

Such low levels of genetic monogamy have surprised biologists and zoologists, forcing them to rethink the role of social monogamy in evolution. They can no longer assume social monogamy determines how genes are distributed in a species. The lower the rates of genetic monogamy among socially monogamous pairs, the less of a role social monogamy plays in determining how genes are distributed among offspring.

Polygyny

Polygyny occurs when one male gets exclusive mating rights with multiple females. In some species, notably those with harem-like structures, only one of a few males in a group of females will mate. Technically, polygyny in sociobiology and zoology is defined as a system in which a male has a relationship with more than one female, but the females are predominantly bonded to a single male. Should the active male be driven out, killed, or otherwise removed from the group, in a number of species the new male will ensure that breeding resources are not wasted on another male's young. The new male may achieve this in many different ways, including:

- Competitive infanticide: in lions, hippopotamuses, and some monkeys, the new male will kill the offspring of the previous alpha male to cause their mothers to become receptive to his sexual advances since they are no longer nursing. To prevent this, many female primates exhibit ovulation cues among all males, and show situation-dependent receptivity.

- Harassment to miscarriage: amongst wild horses and baboons, the male will "systematically harass" pregnant females until they miscarry.

- Pheromone-based spontaneous abortion.

- In some rodents such as mice, a new male with a different scent will cause females who are pregnant to spontaneously fail to implant recently fertilised eggs. This does not require contact; it is mediated by scent alone. It is known as the Bruce effect.

Von Haartman specifically described the mating Behavior of the European pied flycatcher as successive polygyny. Within this system, the males leave their home territory once their primary female lays her first egg. Males then create a second territory, presumably in order to attract a secondary female to breed. Even when they succeed at acquiring a second mate, the males typically return to the first female to exclusively provide for her and her offspring.

Polygynous mating structures are estimated to occur in up to 90% of mammal species. As polygyny is the most common form of polygamy among vertebrates (including humans, to some extent), it has been studied far more extensively than polyandry or polygynandry.

Polyandry

Polyandry occurs when one female gets exclusive mating rights with multiple males. In some species, such as redlip blennies, both polygyny and polyandry are observed.

The anglerfish Haplophryne mollis is polyandrous. This female is trailing
the atrophied remains of males she has encountered.

The males in some deep sea anglerfishes are much smaller than the females. When they find a female they bite into her skin, releasing an enzyme that digests the skin of their mouth and her body and fusing the pair down to the blood-vessel level. The male then slowly atrophies, losing first his digestive organs, then his brain, heart, and eyes, ending as nothing more than a pair of gonads, which release sperm in response to hormones in the female's bloodstream indicating egg release. This extreme sexual dimorphism ensures that, when the female is ready to spawn, she has a mate immediately available. A single anglerfish female can "mate" with many males in this manner.

Polygynandry

Polygynandry occurs when multiple males mate indiscriminately with multiple females. The numbers of males and females need not be equal, and in vertebrate species studied so far, there are usually fewer males. Two examples of systems in primates are promiscuous mating chimpanzees and bonobos. These species live in social groups consisting of several males and several females. Each female copulates with many males, and vice versa. In bonobos, the amount of promiscuity is particularly striking because bonobos use sex to alleviate social conflict as well as to reproduce. This mutual promiscuity is the approach most commonly used by spawning animals, and is perhaps the "original fish mating system." Common examples are forage fish, such as herrings, which form huge mating shoals in shallow water. The water becomes milky with sperm and the bottom is draped with millions of fertilised eggs.

Polygamy

The term polygamy is an umbrella term used to refer generally to non-monogamous matings. As such, polygamous relationships can be polygynous, polyandrous or polygynandrous. In a small number of species, individuals can display either polygamous or monogamous Behavior depending on environmental conditions. An example is the social wasp Apoica flavissima. In some species, polygyny and polyandry is displayed by both sexes in the population. Polygamy in both sexes has been observed in red flour beetle (Tribolium castaneum). Polygamy is also seen in many Lepidoptera species including Mythimna unipuncta (true armyworm moth).

A tournament species is one in which "mating tends to be highly polygamous and involves high levels of male-male aggression and competition." Tournament Behavior often correlates with high levels of sexual dimorphism, examples of species including chimpanzees and baboons. Most polygamous species present high levels of tournament Behavior, with a notable exception being bonobos .

Parental Investment and Reproductive Success

Female and male sexual Behavior differ in many species. Often, males are more active in initiating mating, and bear the more conspicuous sexual ornamentation like antlers and colourful plumage. This is a result of anisogamy, where sperm are smaller and much less costly (energetically) to produce than eggs. This difference in physiological cost means that males are more limited by the number of mates they can secure, while females are limited by the quality of genes of her mates, a phenomenon known as Bateman's principle. Many females also have extra reproductive burdens in that parental care often falls mainly, or exclusively, on them. Thus, females are more limited in their potential reproductive success. In species where males take on more of the reproductive costs, such as sea horses and jacanas, the role is reversed, and the females are larger, more aggressive and more brightly coloured than the males.

In hermaphroditic animals, the costs of parental care can be evenly distributed between the sexes, e.g. earthworms. In some species of planarians, sexual Behavior takes the form of penis fencing. In this form of copulation, the individual that first penetrates the other with the penis, forces the other to be female, thus carrying the majority of the cost of reproduction. Post mating, banana slugs will some times gnaw off their partners penis as an act of sperm competition called apophallation. This is costly as they must heal, and spend more energy courting conspecifics that can act as male and female. A hypothesis suggests these slugs may be able to compensate the loss of the male function by directing energy that would have been put towards it to the female function. In the grey slug, the sharing of cost leads to a spectacular display, where the mates suspend themselves high above the ground from a slime thread, ensuring none of them can refrain from taking on the cost of egg-bearer.

Mating grey slugs, suspended from a slime thread.

Seasonality

Many animal species have specific mating (or breeding) periods e.g. (seasonal breeding) so that offspring are born or hatch at an optimal time. In marine species with limited mobility and external fertilisation like corals, sea urchins and clams, the timing of the common spawning is the only externally visible form of sexual Behavior. In areas with continuously high primary production, some species have a series of breeding seasons throughout the year. This is the case with most primates (who are primarily tropical and subtropical animals). Some animals (opportunistic breeders) breed dependent upon other conditions in their environment aside from time of year.

Mammals

Mating seasons are often associated with changes to herd or group structure, and Behavioral changes, including territorialism amongst individuals. These may be annual (e.g. wolves), biannual (e.g. dogs) or more frequently (e.g. horses). During these periods, females of most mammalian species are more mentally and physically receptive to sexual advances, a period scientifically described as estrous but commonly described as being "in season" or "in heat". Sexual Behavior may occur outside estrus, and such acts as do occur are not necessarily harmful.

Some mammals (e.g. domestic cats, rabbits and camilidae) are termed "induced ovulators". For these species, the female ovulates due to an external stimulus during, or just prior, to mating, rather than ovulating cyclically or spontaneously. Stimuli causing induced ovulation include the sexual Behavior of coitus, sperm and pheromones. Domestic cats have penile spines. Upon withdrawal of a cat's penis, the spines rake the walls of the female's vagina, which may cause ovulation.

Amphibians

For many amphibians, an annual breeding cycle applies, typically regulated by ambient temperature, precipitation, availability of surface water and food supply. This breeding season is accentuated in temperate regions, in boreal climate the breeding season is typically concentrated to a few short days in the spring. Some species, such as the Rana Clamitans (green frog), spend from June to August defending their territory. In order to protect these territories, they use five vocalizations.

Fish

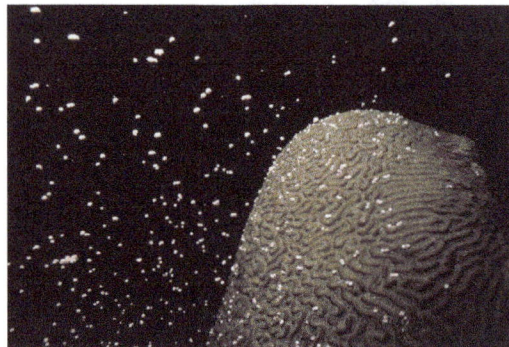

Brain corals typically spawning in connection with the full moon every August.

Like many coral reef dwellers, the clownfish spawn around the time of the full moon in the wild. In a group of clownfish, there is a strict dominance hierarchy. The largest and most aggressive female is found at the top. Only two clownfish, a male and a female, in a group reproduce through external fertilisation. Clownfish are sequential hermaphrodites, meaning that they develop into males first, and when they mature, they become females. If the female clownfish is removed from the group, such as by death, one of the largest and most dominant males will become a female. The remaining males will move up a rank in the hierarchy.

Motivation

Various neurohormones stimulate sexual wanting in animals. In general, studies have suggested that dopamine is involved in sexual incentive motivation, oxytocin and melanocortins in sexual attraction, and noradrenaline in sexual arousal. Vasopressin is also involved in the sexual Behavior of some animals.

Neurohormones in the Mating Systems of Voles

The mating system of prairie voles is monogamous; after mating, they form a lifelong bond. In contrast, montane voles have a polygamous mating system. When montane voles mate, they form no strong attachments, and separate after copulation. Studies on the brains of these two species have found that it is two neurohormones and their respective receptors that are responsible for these differences in mating strategies. Male prairie voles release vasopressin after copulation with a partner, and an attachment to their partner then develops. Female prairie voles release oxytocin after copulation with a partner, and similarly develop an attachment to their partner.

Neither male nor female montane voles release high quantities of oxytocin or vasopressin when they mate. Even when injected with these neurohormones, their mating system does not change. In contrast, if prairie voles are injected with the neurohormones, they may form a lifelong attachment, even if they have not mated. It's believed that the differing response to the neurohormones between the two species is due to a difference in the number of oxytocin and vasopressin receptors. Prairie voles have a greater number of oxytocin and vasopressin receptors compared to montane voles, and are therefore more sensitive to those two neurohormones. It's believed that it's the quantity of receptors, rather than the quantity of the hormones, that determines the mating system and bond-formation of either species.

Oxytocin and Rat Sexual Behavior

Mother rats experience a postparum estrus which makes them highly motivated to mate. However, they also have a strong motivation to protect their newly born pups. As a consequence, the mother rat solicits males to the nest but simultaneously becomes aggressive towards them to protect her young. If the mother rat is given injections of an oxytocin receptor antagonist, they no longer experience these maternal motivations.

Prolactin influences social bonding in rats.

Oxytocin and Primate Sexual Behavior

Oxytocin plays a similar role in non-human primates as it does in humans.

Grooming, sex, and cuddling frequencies correlate positively with levels of oxytocin. As the level of oxytocin increases so does sexual motivation. While oxytocin plays a major role in parent child relationships, it is also found to play a role in adult sexual relationships. Its secretion affects the nature of the relationship or if there will even be a relationship at all.

Studies have shown that oxytocin is higher in monkeys in lifelong monogamous relationships compared to monkeys which are single. Furthermore, the oxytocin levels of the couples correlate positively; when the oxytocin secretion of one increases the other one also increases. Higher levels of oxytocin are related to monkeys expressing more Behaviors such as cuddling, grooming and sex, while lower levels of oxytocin reduce motivation for these activities.

Research on oxytocin's role in the animal brain suggests that it plays less of a role in Behaviors of love and affection than previously believed. "When oxytocin was first discovered in 1909, it was thought mostly to influence a mother's labour contractions and milk let-down. Then, in the 1990s, research with prairie voles found that giving them a dose of oxytocin resulted in the formation of a bond with their future mate." Oxytocin has since been treated by the media as the sole player in the "love and mating game" in mammals. This view, however, is proving to be false as, "most hormones don't influence Behavior directly. Rather, they affect thinking and emotions in variable ways." There is much more involved in sexual Behavior in the mammalian animal than oxytocin and vasopressin can explain.

Pleasure

It is often assumed that animals do not have sex for pleasure, or alternatively that humans, pigs, bonobos (and perhaps dolphins and one or two more species of primates) are the only species that do. This is sometimes stated as "animals mate only for reproduction". This view is considered a misconception by some scholars. Jonathan Balcombe argues that the prevalence of non-reproductive sexual Behavior in certain species suggests that sexual stimulation is pleasurable. He also points to the presence of the clitoris in some female mammals, and evidence for female orgasm in primates. On the other hand, it is impossible to know the subjective feelings of animals, and the notion that non-human animals experience emotions similar to humans is a contentious subject.

A 2006 Danish Animal Ethics Council report, which examined current knowledge of animal sexuality in the context of legal queries concerning sexual acts by humans, has the following comments, primarily related to domestically common animals:

> "Even though the evolution-related purpose of mating can be said to be reproduction, it is not actually the creating of offspring which originally causes them to mate. It is probable that they mate because they are motivated for the actual copulation, and because this is connected with a positive experience. It is therefore reasonable to assume that there is some form of pleasure or satisfaction connected with the act. This assumption is confirmed by the Behavior of males, who in the case of many species are prepared to work to get access to female animals, especially if the female animal is in oestrus, and males who for breeding purposes are used to having sperm collected become very eager, when the equipment they associate with the collection is taken out."

There is nothing in female mammals' anatomy or physiology that contradicts that stimulation of the sexual organs and mating is able to be a positive experience. For instance, the clitoris acts in the same way as with women, and scientific studies have shown that the success of reproduction is improved by stimulation of clitoris on (among other species) cows and mares in connection with insemination, because it improves the transportation of the sperm due to contractions of the inner genitalia. This probably also applies to female animals of other animal species, and contractions in the inner genitals are seen e.g. also during orgasm for women. It is therefore reasonable to assume that sexual intercourse may be linked with a positive experience for female animals.

Koinophilia

Koinophilia is the love of the "normal" or phenotypically common. The term was introduced to scientific literature in 1990, and refers to the tendency of animals seeking a mate to prefer that mate not to have any unusual, peculiar or deviant features. Similarly, animals preferentially choose mates with low fluctuating asymmetry. However, animal sexual ornaments can evolve through runaway selection, which is driven by (usually female) selection for non-standard traits.

Interpretation Bias

The field of study of sexuality in non-human species was a long-standing taboo. In the past, researchers sometimes failed to observe, mis-categorising and mis-described sexual Behavior which did not meet their preconceptions (mainly from western puritanism). In earlier periods, bias tended to support what would now be described as conservative sexual mores. An example of overlooking Behavior relates to descriptions of giraffe mating:

> When nine out of ten pairings occur between males, "every male that sniffed a female was reported as sex, while anal intercourse with orgasm between males was only categorized as 'revolving around' dominance, competition or greetings."

In the 21st century, liberal social or sexual views are often projected upon animal subjects of research. Popular discussions of bonobos are a frequently cited example. Current research frequently expresses views such as that of the Natural History Museum at the University of Oslo, which in 2006 held an exhibition on animal sexuality:

> "Many researchers have described homosexuality as something altogether different from sex. They must realise that animals can have sex with who they will, when they will and without consideration to a researcher's ethical principles."

Other animal activities may be misinterpreted due to the frequency and context in which animals perform the Behavior. For example, domestic ruminants display Behaviors such as mounting and head-butting. This often occurs when the animals are establishing dominance relationships and are not necessarily sexually motivated. Careful analysis must be made to interpret what animal motivations are being expressed by those Behaviors.

Types of Sexual Behavior

Copulation

Copulation is the union of the male and female sex organs, the innate sexual activity specifically organized to transmit male sperm into the body of the female.

In non-primate mammals (for example, rodents, canines, felines, bovines, and equines), the anatomy of the reproductive organs and some circuits of the nervous system are specifically organized for heterosexual copulation.

Cuckoldry

Alternative male strategies which allow small males to engage in cuckoldry can develop in species such as fish where spawning is dominated by large and aggressive males. Cuckoldry is a variant of polyandry, and can occur with sneak spawners. A sneak spawner is a male that rushes in to join the spawning rush of a spawning pair. A spawning rush occurs when a fish makes a burst of speed, usually on a near vertical incline, releasing gametes at the apex, followed by a rapid return to the lake or sea floor or fish aggregation. Sneaking males do not take part in courtship. In salmon and trout, for example, jack males are common. These are small silvery males that migrate upstream along with the standard, large, hook-nosed males and that spawn by sneaking into redds to release sperm simultaneously with a mated pair. This Behavior is an evolutionarily stable strategy for reproduction, because it is favoured by natural selection just like the "standard" strategy of large males.

Small male bluegill sunfishes cuckold large males by adopting sneaker strategies.

Hermaphroditism

Hermaphroditism occurs when a given individual in a species possesses both male and female reproductive organs, or can alternate between possessing first one, and then the other. Hermaphroditism is common in invertebrates but rare in vertebrates. It can be contrasted with gonochorism, where each individual in a species is either male or female, and remains that way throughout their lives. Most fish are gonochorists, but hermaphroditism is known to occur in 14 families of teleost fishes.

Female groupers change their sex to male if no male is available.

Usually hermaphrodites are sequential, meaning they can switch sex, usually from female to male (protogyny). This can happen if a dominant male is removed from a group of females. The largest female in the harem can switch sex over a few days and replace the dominant male. This is found amongst coral reef fishes such as groupers, parrotfishes and wrasses. It is less common for a male to switch to a female (protandry). As an example, most wrasses are protogynous hermaphrodites within a haremic mating system. Hermaphroditism allows for complex mating systems. Wrasses exhibit three different mating systems: polygynous, lek-like, and promiscuous mating systems.

Sexual Cannibalism

Sexual cannibalism is a Behavior in which a female animal kills and consumes the male before, during, or after copulation. Sexual cannibalism confers fitness advantages to both the male and female. Sexual cannibalism is common among insects, arachnids and amphipods. There is also evidence of sexual cannibalism in gastropods and copepods.

Sexual Coercion

Sex in a forceful or apparently coercive context has been documented in a variety of species. In some herbivorous herd species, or species where males and females are very different in size, the male dominates sexually by force and size.

Some species of birds have been observed combining sexual intercourse with apparent violent assault; these include ducks, and geese. Female white-fronted bee-eaters are subjected to forced copulations. When females emerge from their nest burrows, males sometimes force them to the ground and mate with them. Such forced copulations are made preferentially on females who are laying and who may therefore lay eggs fertilized by the male.

It has been reported that young male elephants in South Africa sexually coerced and killed rhinoceroses. This interpretation of the elephants' Behavior was disputed by one of the original study's authors, who said there was "nothing sexual about these attacks".

During mating, the male muscovy duck typically immobilises the female.

Parthenogenesis

Parthenogenesis is a form of asexual reproduction in which growth and development of embryos occur without fertilisation. Technically, parthenogenesis is not a Behavior, however, sexual Behaviors may be involved.

Whip-tailed lizard females have the ability to reproduce through parthenogenesis and as such males are rare and sexual breeding non-standard. Females engage in "pseudocopulation" to stimulate ovulation, with their Behavior following their hormonal cycles; during low levels of oestrogen, these (female) lizards engage in "masculine" sexual roles. Those animals with currently high oestrogen levels assume "feminine" sexual roles. Lizards that perform the courtship ritual have greater fecundity than those kept in isolation due to an increase in hormones triggered by the sexual Behaviors. So, even though asexual whiptail lizards populations lack males, sexual stimuli still increase reproductive success. From an evolutionary standpoint these females are passing their full genetic code to all of their offspring rather than the 50% of genes that would be passed in sexual reproduction.

It is rare to find true parthenogenesis in fishes, where females produce female offspring with no input from males. All-female species include the Texas silverside, Menidia clarkhubbsi as well as a complex of Mexican mollies.

Parthenogenesis has been recorded in 70 vertebrate species including hammerhead sharks, blacktip sharks, amphibians and crayfish.

Unisexuality

Unisexuality occurs when a species is all-male or all-female. Unisexuality occurs in some fish species, and can take complex forms. Squalius alburnoides, a minnow found in several river basins in Portugal and Spain, appears to be an all-male species. The existence of this species illustrates the potential complexity of mating systems in fish. The species originated as a hybrid between two species, and is diploid, but not hermaphroditic. It can have triploid and tetraploid forms, including all-female forms that reproduce mainly through hybridogenesis.

- Interbreeding: Hybrid offspring can result from the mating of two organisms of distinct but closely related parent species, although the resulting offspring is not always fertile. According to Alfred Kinsey, genetic studies on wild animal populations have shown a "large number" of inter-species hybrids.

- Prostitution: There are reports that animals occasionally engage in prostitution. A small number of pair-bonded females within a group of penguins took nesting material (stones) after copulating with a non-partner male.

- Pavlovian conditioning: The sexualisation of objects or locations is recognised in the animal breeding world. For example, male animals may become sexually aroused upon visiting a location where they have been allowed to have sex before, or upon seeing a stimulus previously associated with sexual activity such as an artificial vagina. Sexual preferences for certain cues can be artificially induced in rats by pairing scents or objects with their early sexual experiences. The primary motivation of this Behavior is Pavlovian conditioning, and the association is due to a conditioned response (or association) formed with a distinctive "reward".

- Viewing images: A study using four adult male rhesus macaques (Macaca mulatta) showed that male rhesus macaques will give up a highly valued item, juice. Encouraging captive pandas to mate is problematic. Showing young male pandas "panda pornography" is credited with a recent population boom among pandas in captivity in China. One researcher attributed the success to the sounds on the recordings.

- Copulatory wounding and Traumatic Insemination: Injury to a partner's genital tract during mating occurs in at least 40 taxa, ranging from fruit flies to humans. However, it often goes unnoticed due to its cryptic nature and because of internal wounds not visible outside.

A dog mates with a coyote to produce a dog-coyote hybrid.

Non-reproductive Sexual Behavior

There is a range of Behaviors that animals perform which appear to be sexually motivated but which can not result in reproduction. These include:

- Masturbation: Some species, both male and female, masturbate, both when partners are available and otherwise.

- Oral sex: Several species engage in both autofellatio and oral sex. This has been documented in brown bears, Tibetan macaques, wolves, goats, primates, hyenas, bats, cape ground squirrels and sheep. In the greater short-nosed fruit bat, copulation by males is dorsoventral and the females lick the shaft or the base of the male's penis, but not the glans which has already penetrated the vagina. While the females do this, the penis is not withdrawn and research has shown a positive relationship between length of the time that the penis is licked and the duration of copulation. Post copulation genital grooming has also been observed.

- Homosexuality: Same-sex sexual Behavior occurs in a range of species, especially in social species, particularly in marine birds and mammals, monkeys, and the great apes. As of 1999, the scientific literature contained reports of homosexual Behavior in at least 471 wild species. Organisers of the Against Nature? exhibit stated that "homosexuality has been observed among 1,500 species, and that in 500 of those it is well documented."

- Genital-genital rubbing: This is sexual activity in which one animal rubs his or her genitals against the genitals of another animal. This is stated to be the "bonobo's most typical sexual pattern, undocumented in any other primate".

- Inter-species mating: Some animals opportunistically mate with individuals of another species.

- Sex involving juveniles: Male stoats (Mustela erminea) will sometimes mate with infant females of their species. This is a natural part of their reproductive biology – they have a delayed gestation period, so these females give birth the following year when they are fully grown. Juvenile male chimpanzees have been recorded mounting and copulating with immature chimps. Infants in bonobo societies are often involved in sexual Behavior.

- Necrophilia: This describes when an animal engages in a sexual act with a dead animal. It has been observed in mammals, birds, reptiles and frogs.

- Bisexuality: This describes when an animal shows sexual Behavior towards both males and females.

- Extended female sexuality: This is when females mate with males outside of their conceptive period.

A male black and white tegu mounts a female that has been
dead for two days and attempts to mate.

Seahorse

Seahorses, once considered to be monogamous species with pairs mating for life, were described in a 2007 study as "promiscuous, flighty, and more than a little bit gay". Scientists at 15 aquaria studied 90 seahorses of three species. Of 3,168 sexual encounters, 37% were same-sex acts. Flirting was common (up to 25 potential partners a day of both sexes); only one species (the British spiny seahorse) included faithful representatives, and for these 5 of 17 were faithful, 12 were not. Bisexual Behavior was widespread and considered "both a great surprise and a shock", with big-bellied seahorses of both sexes not showing partner preference. 1,986 contacts were male-female, 836 were female-female and 346 were male-male.

Bonobo

The bonobo is a fully bisexual species. Both males and females engage in sexual Behavior with the same and the opposite sex, with females being particularly noted for engaging in sexual Behavior with each other and at up to 75% of sexual activity being nonreproductive, as being sexually active does not necessarily correlate with their ovulation cycles. Sexual activity occurs between almost all ages and sexes of bonobo societies. Primatologist Frans de Waal believes that bonobos use sexual activity to resolve conflict between individuals. Immature bonobos, contrariwise, perform genital contact when relaxed.

Bonobos mating, Jacksonville Zoo and Gardens.

Macaque

Similar same-sex sexual Behaviors occur in both male and female macaques. It is thought to be done for pleasure as an erect male mounts and thrusts upon or into another male. Sexual receptivity can also be indicated by red faces and shrieking. Mutual ejaculation after a combination of anal intercourse and masturbation has also been witnessed, although it may be rare. In comparison to socio-sexual Behaviors such as dominance displays, homosexual mounts last longer, happen in series, and usually involves pelvic thrusting.

Females are also thought to participate for pleasure as VPA (vulvar, perineal, and anal) stimulation is part of these interactions. The stimulation can come from their own tails, mounting their partner, thrusting, contact between both VPAs, or a combination of these.

Dolphin

Male bottlenose dolphins have been observed working in pairs to follow or restrict the movement of a female for weeks at a time, waiting for her to become sexually receptive. The same pairs have also been observed engaging in intense sexual play with each other. Janet Mann, a professor of biology and psychology at Georgetown University, argues that the common same-sex Behavior among male dolphin calves is about bond formation and benefits the species evolutionarily. They cite studies that have shown the dolphins later in life are bisexual and the male bonds forged from homosexuality work for protection as well as locating females with which to reproduce. In 1991, an English man was prosecuted for allegedly having sexual contact with a dolphin. The man was found not guilty after it was revealed at trial that the dolphin was known to tow bathers through the water by hooking his penis around them.

Hyena

The female spotted hyena has a unique urinary-genital system, closely resembling the penis of the male, called a pseudo-penis. Dominance relationships with strong sexual elements are routinely observed between related females. They are notable for using visible sexual arousal as a sign of submission but not dominance in males as well as females (females have a sizable erectile clitoris). It is speculated that to facilitate this, their sympathetic and parasympathetic nervous systems may be partially reversed in respect to their reproductive organs.

Mating Behavior

Mammals

Mammals mate by vaginal copulation. To achieve this, the male usually mounts the female from behind. The female may exhibit lordosis in which she arches her back ventrally to facilitate entry of the penis. Amongst the land mammals, other than humans, only bonobos mate in a face-to-face position, as the females' anatomy seems to reflect, although ventro-ventral copulation has also been observed in Rhabdomys. Some sea mammals copulate in a belly-to-belly position. Some camelids mate in a lying-down position. In most mammals ejaculation occurs after multiple intromissions, but in most primates, copulation consists of one brief intromission. In most ruminant species, a single pelvic thrust occurs during copulation. In most deer species, a copulatory jump also occurs.

During mating, a "copulatory tie" occurs in mammals such as fossas, canids and Japanese martens. A "copulatory lock" also occurs in some primate species, such as Galago senegalensis.

The copulatory behavior of many mammalian species is affected by sperm competition.

Some females have concealed fertility, making it difficult for males to evaluate if a female is fertile. This is costly as ejaculation expends much energy.

Invertebrates

Invertebrates are often hermaphrodites. Some hermaphroditic land snails begin mating with an elaborate tactile courting ritual. The two snails circle around each other for up to six hours, touching with their tentacles, and biting lips and the area of the genital pore, which shows some preliminary signs of the eversion of the penis. As the snails approach mating, hydraulic pressure builds up in the blood sinus surrounding an organ housing a sharpened dart. The dart is made of calcium carbonate or chitin, and is called a love dart. Each snail manoeuvres to get its genital pore in the best position, close to the other snail's body. Then, when the body of one snail touches the other snail's genital pore, it triggers the firing of the love dart. After the snails have fired their darts, they copulate and exchange sperm as a separate part of the mating progression. The love darts are covered with a mucus that contains a hormone-like substance that facilitates the survival of the sperm.

Penis fencing is a mating Behavior engaged in by certain species of flatworm, such as Pseudobiceros bedfordi. Species which engage in the practice are hermaphroditic, possessing both eggs and sperm-producing testes. The species "fence" using two-headed dagger-like penises which are pointed, and white in colour. One organism inseminates the other. The sperm is absorbed through pores in the skin, causing fertilisation.

Corals can be both gonochoristic (unisexual) and hermaphroditic, each of which can reproduce sexually and asexually. Reproduction also allows corals to settle new areas. Corals predominantly reproduce sexually. 25% of hermatypic corals (stony corals) form single sex (gonochoristic) colonies, while the rest are hermaphroditic. About 75% of all hermatypic corals "broadcast spawn" by releasing gametes – eggs and sperm – into the water to spread offspring. The gametes fuse during fertilisation to form a microscopic larva called a planula, typically pink and elliptical in shape. Synchronous spawning is very typical on the coral reef and often, even when multiple species are present, all corals spawn on the same night. This synchrony is essential so that male and female gametes can meet. Corals must rely on environmental cues, varying from species to species, to determine the proper time to release gametes into the water. The cues involve lunar changes, sunset time, and possibly chemical signalling. Synchronous spawning may form hybrids and is perhaps involved in coral speciation.

Courting garden snails: The one on the left has
fired a love dart into the one on the right.

Butterflies spend much time searching for mates. When the male spots a mate, he will fly closer and release pheromones. He then performs a special courtship dance to attract the female. If the female appreciates the dancing she may join him. Then they join their bodies together end to end at their abdomens. Here, the male passes the sperm to the female's egg-laying tube, which will soon be fertilised by the sperm.

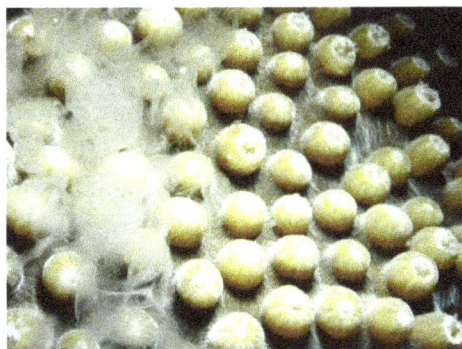

A male star coral releases sperm into the water.

Many animals make plugs of mucus to seal the female's orifice after mating. Normally such plugs are secreted by the male, to block subsequent partners. In spiders the female can assist the process. Spider sex is unusual in that males transfer their sperm to the female on small limbs called pedipalps.

They use these to pick their sperm up from their genitals and insert it into the female's sexual orifice, rather than copulating directly. On the 14 occasions a sexual plug was made, the female produced it without assistance from the male. On ten of these occasions the male's pedipalps then seemed to get stuck while he was transferring the sperm (which is rarely the case in other species of spider), and he had great difficulty freeing himself. In two of those ten instances, he was eaten as a result.

Genetic Evidence of Interspecies Sexual Activity in Humans

Research into human evolution confirms that, in some cases, interspecies sexual activity may have been responsible for the evolution of new species (speciation). Analysis of animal genes found evidence that after humans had diverged from other apes, interspecies mating nonetheless occurred regularly enough to change certain genes in the new gene pool. Researchers found that the X chromosomes of humans and chimps may have diverged around 1.2 million years after the other chromosomes. One possible explanation is that modern humans emerged from a hybrid of human and chimp populations. A 2012 study questioned this explanation, concluding that "there is no strong reason to involve complicated factors in explaining the autosomal data".

Inbreeding Avoidance

When close relatives mate, progeny may exhibit the detrimental effects of inbreeding depression. Inbreeding depression is predominantly caused by the homozygous expression of recessive deleterious alleles. Over time, inbreeding depression may lead to the evolution of inbreeding avoidance Behavior. Several examples of animal Behavior that reduce mating of close relatives and inbreeding depression are described next.

Reproductively active female naked mole-rats tend to associate with unfamiliar males (usually non-kin), whereas reproductively inactive females do not discriminate. The preference of reproductively active females for unfamiliar males is interpreted as an adaptation for avoiding inbreeding.

When mice inbreed with close relatives in their natural habitat, there is a significant detrimental effect on progeny survival. In the house mouse, the major urinary protein (MUP) gene cluster provides a highly polymorphic scent signal of genetic identity that appears to underlie kin recognition and inbreeding avoidance. Thus there are fewer matings between mice sharing MUP haplotypes than would be expected if there were random mating.

Meerkat females appear to be able to discriminate the odour of their kin from the odour of their non-kin. Kin recognition is a useful ability that facilitates both cooperation among relatives and the avoidance of inbreeding. When mating does occur between meerkat relatives, it often results in inbreeding depression. Inbreeding depression was evident for a variety of traits: pup mass at emergence from the natal burrow, hind-foot length, growth until independence and juvenile survival.

The grey-sided vole (Myodes rufocanus) exhibits male-biased dispersal as a means of avoiding incestuous matings. Among those matings that do involve inbreeding the number of weaned juveniles in litters is significantly smaller than that from non-inbred litters indicating inbreeding depression.

In natural populations of the bird Parus major (great tit), inbreeding is likely avoided by dispersal of individuals from their birthplace, which reduces the chance of mating with a close relative.

Toads display breeding site fidelity, as do many amphibians. Individuals that return to natal ponds to breed will likely encounter siblings as potential mates. Although incest is possible, Bufo americanus siblings rarely mate. These toads likely recognise and actively avoid close kins as mates. Advertisement vocalisations by males appear to serve as cues by which females recognise their kin.

ADJUNCTIVE BEHAVIOR

Adjunctive Behavior occurs when an animal expresses an activity reliably accompanying some other response that has been produced by a stimulus, especially when the stimulus is presented according to a temporally defined schedule. For example, in 1960, psychologist John Falk was studying hungry rats that had been trained to press a lever for a small food pellet. Once a rat had received a pellet, it was obliged to wait an average of one minute before another press of the lever would be rewarded. The rats developed the habit of drinking water during these intervals, but their consumption far exceeded what was expected. Many consumed three to four times their normal daily water intake during a three-hour session, and some drank nearly half of their body weight in water during this time. Further research has revealed that intermittent food presentation to a variety of organisms results in an inordinately excessive consumption of water as well as other Behaviors including attack, pica, escape, and alcohol consumption.

In psychological terminology, adjunctive Behavior is non-contingent Behavior maintained by an event which acquires a reinforcing effect due to some other reinforcing contingency. Some usages emphasize the stimulus rather than the responding it engenders (e.g., in rats, food presentations typically produce eating reliably followed by drinking; the drinking is adjunctive and is sometimes said to be induced by the schedule of food presentation).

Use in Science

Adjunctive Behavior has been used as evidence of animal welfare problems. Pregnant sows are typically fed only a fraction of the amount of food they would consume by choice, and they remain hungry for almost the whole day. If a water dispenser is available, some sows will drink two or three times their normal daily intake, and under winter conditions, warming this amount of cold water to body temperature, only to discharge it as dilute urine, involves an appreciable caloric cost. However, if such sows are given a bulky high-fibre food (which under typical circumstances would result in an increase in water intake), they spend much longer eating, and the excessive drinking largely disappears. In this case, much of the sows' water intake appeared to be adjunctive drinking that was not linked to thirst.

ABNORMAL BEHAVIORS IN ANIMALS

Abnormal Behavior in animals can be defined in several ways. Statistically, abnormal is when the occurrence, frequency or intensity of a Behavior varies statistically significantly, either more or less, from the normal value. This means that theoretically, almost any Behavior could become abnormal in an individual. Less formally, 'abnormal' includes any activity judged to be outside the normal Behavior pattern for animals of that particular class or age. For example, infanticide may be a normal Behavior and regularly observed in one species, however, in another species it might be normal but becomes 'abnormal' if it reaches a high frequency, or in another species it is rarely observed and any incidence is considered 'abnormal'. This list does not include one-time Behaviors performed by individual animals that might be considered abnormal for that individual, unless these are performed repeatedly by other individuals in the species and are recognised as part of the ethogram of that species.

Most abnormal Behaviors can be categorised collectively (e.g., eliminative, ingestive, stereotypies), however, many abnormal Behaviors fall debatedly into several of these categories and categorisation is therefore not attempted in this list. Some abnormal Behaviors may be related to environmental conditions (e.g. captive housing) whereas others may be due to medical conditions.

References

- Animal-behavior, science: britannica.com, Retrieved 14 May, 2019

- Thorpe, Showick; Thorpe, Edgar (2009). General Studies Manual. Pearson Education India. P. 17. ISBN 9788131721339

- Ethology-definition-and-approaches-zoology, ethology: notesonzoology.com, Retrieved 15 June, 2019

- Westneat, D. F.; Stewart, I. R. K. (2003). "EXTRA-PAIRPATERNITY INBIRDS: Causes, Correlates, and Conflict". Annual Review of Ecology, Evolution, and Systematics. 34: 365–396. Doi:10.1146/annurev.ecolsys.34.011802.132439

- Neuroethology: scholarpedia.org, Retrieved 16 July, 2019

- Blackshaw, J.K. "Behavioural profiles of domestic animals - horses". Archived from the original on March 17, 2012. Retrieved March 5, 2013

- Innate-Behavior-of-Animals-BIO, lesson, innate-behavior, biology: ck12.org, Retrieved 17 August, 2019

- Baker, K., Bloomsmith, M., Griffis, C. And Gierhart, M., (2003). Self injurious behavior and response to human interaction as enrichment in rhesus macaques" American Journal of Primatology 60 (Suppl. 1): 94-95. ISSN 0275-2565

- Section , CK-12-Biology-Advanced-Concepts: ck12.org, Retrieved 18 January, 2019

- Animal-social-behaviour, topic: britannica.com, Retrieved 19 February, 2019

Permissions

Index